Advanced Environmental Wind Engineering

Yukio Tamura • Ryuichiro Yoshie
Editors

Advanced Environmental Wind Engineering

Springer

Editors
Yukio Tamura
Tokyo Polytechnic University
Kanagawa
Japan

Beijing Jiaotong University
Beijing
China

Ryuichiro Yoshie
Tokyo Polytechnic University
Kanagawa
Japan

ISBN 978-4-431-56727-1 ISBN 978-4-431-55912-2 (eBook)
DOI 10.1007/978-4-431-55912-2

Printed on acid-free paper

This Springer imprint is published by Springer Nature
The registered company is Springer Japan KK

Preface

The effect of wind is a very important issue in the building and structural engineering field. It is said that almost 70–80 % of economic losses due to natural disasters in the world are caused by extreme winds and related water hazards. The risk of future disasters continues to escalate with population shifts towards urban centers and the impending threat of their increased intensity and frequency as hypothesized by potential climate change. Urbanization has also led to the deterioration of regional and global environmental quality with a far-reaching impact on public health. Not only extremely strong wind events or other natural hazards of short duration, but also serious environmental problems should be treated as long-lasting hazards causing severe disasters in human society. This calls for a sustainable society that emphasizes reduced energy consumption and improved environmental quality.

The over-arching vision of the Tokyo Polytechnic University-Wind engineering Group (TPU–WEG) is to build a sustainable urban environment that is resilient to wind-related events and is in harmony with regional local climates. Its focus is on developing an integrated education and research program that covers a wide spectrum of problems to address wind-related challenges of the next frontiers in urban regions of Asia and beyond.

In order to create efficient and comfortable environments, and to satisfy people's basic demand for iconic symbols in cities, buildings are becoming taller and slender, and bridges and roof structures are becoming longer and lighter, thus making them more vulnerable to wind. The importance of wind engineering is significantly increasing, but few universities have curriculums in this field.

Since July 2006, the TPU–WEG has been organizing the International Advanced School (IAS) on Wind Engineering with the intention of covering this lack of relevant advanced professional training.

Thirty-seven world-eminent professors and researchers were invited as lecturers to the past nine IAS sessions: Chris Baker, Shuyang Cao, Qingyan Chen, Chiiming Cheng, Tadeusz Chmielewski, Richard de Dear, David Etheridge, Richard Flay, Yaojun Ge, Jamie Hernandez, John Holmes, Ahsan Kareem, Michael Kasperski,

Shinsuke Kato, Youngduk Kim, Greg Kopp, Prem Krishna, Takashi Kurabuchi, Kenny Kwok, Sangjoon Lee, Chris Letchford, Xiaofeng Li, Rheo Lim, Masaru Matsumoto, Robert Meroney, Akashi Mochida, William Nazaroff, Siva Parameswaran, Michael Reyes, Matthew Santamouris, Partha Sarkar, Michael Schatzmann, Giovanni Solari, Ted Stathopoulos, Charles Weschler, Youlin Xu, and Lingmi Zhang. They collaborate with three lecturers from TPU–WEG—Masaaki Ohba, Ryuichiro Yoshie, and me – to make the IAS sessions fruitful and successful.

The IAS lectures dealt with wind climates, wind-induced disasters, structural aerodynamics, wind loads, and various wind environmental problems. The lectures provided necessary basic knowledge as well as the latest, state-of-the-art information. The IAS also provided a platform for exchanging and sharing information through discussions between leading researchers in the world and young people new to the field, so that serious wind-related problems regarding wind hazard risk due to strong winds, urban air pollution, and increase of environmental load will be solved by young leaders in the future. On behalf of TPU–WEG and the TPU Global Center of Excellence (COE) program members, I would like to express our sincere gratitude to all lecturers and participants for their great contributions and achievements.

Nine lecture notes in the environmental wind engineering field prepared for the past IAS sessions were rearranged for publication in this book, *Advanced Environmental Wind Engineering*. A related volume titled *Advanced Structural Wind Engineering* based on the IAS lecture notes in the structural wind engineering field was already published in 2013.

I am confident that these two books will be useful to students, engineers, and researchers who work in relevant scientific research or design topics, and will enable them to contribute to the development and construction of sustainable urban environments and hence to achieve wind hazard resilient cities.

Finally, I would like to express my gratitude to all the chapter authors for their cooperation in the creation of these books, providing a valuable contribution to the common goal of the structural and environmental wind engineering.

Kanagawa, Japan
Beijing, China

Yukio Tamura

Contents

Chapter 1
Design Procedures for Natural Ventilation

David Etheridge

Abstract The paper gives an overview of the design process for natural ventilation. It concentrates on the technical procedures that are available to the designer. The overall process is discussed in terms of four stages. For the first stage, the factors that determine the feasibility or otherwise of natural ventilation are outlined. In the second stage, the various ventilation strategies are considered. The remaining two stages involve quantitative design procedures. Prior to describing these procedures, the physical mechanisms of natural ventilation are summarised.

Keywords Natural ventilation • Design • Scale modelling • Theoretical modelling

1.1 Introduction

The aim of the first lecture is to give an overview of available procedures (experimental and theoretical) for natural ventilation design. To put these procedures into context, the whole design process is described in terms of four stages. It is not claimed that this is a definitive description nor that it is a complete one. However it is believed that it covers most of the technical issues that specifically arise with natural ventilation. It is based on the form of Chap. 4 in CIBSE (2005). There are of course other important issues that arise in practice, e.g. client acceptability, cost, detail design of components, diagnostic tools, fire safety and security. Descriptions

Note by Author This chapter is the unchanged text of the first of four related lectures on natural ventilation, given in 2007 at the COE International Advanced School in Korea. Of the three other lectures (Etheridge 2007a, b, c), the third lecture (Etheridge 2007b) is given here as Chap. 2, again in its original form. It should be noted that a much more comprehensive and updated treatment of natural ventilation can now be found in the recent book by the author published in 2012 (Etheridge DW (2012) Natural ventilation of buildings – theory, measurement and design. Wiley, Chichester).

D. Etheridge (✉)
Old Hunstanton, Norfolk PE36 6EQ, UK

previously Department of Architecture and Built Environment, University of Nottingham, Nottingham, UK
e-mail: etheridge.david@btinternet.com

Y. Tamura, R. Yoshie (eds.), *Advanced Environmental Wind Engineering*,
DOI 10.1007/978-4-431-55912-2_1

of such issues can be found in publications relating to specific buildings and in more general works such as CIBSE (2005), Allard (1998) and NATVENT (1997).

1.1.1 Overall Design Process

The design of a natural ventilation system can range from a very simple process to a very complex one. The simplest procedure is to reproduce an existing design that is known to function satisfactorily under given conditions. This procedure has been followed for thousands of years and has led to the development of traditional designs adapted to specific climates. Most single-family houses are probably designed in this way. With non-residential buildings, the process is more demanding, partly because the building is likely to have some unique properties and partly because the requirements themselves are more demanding (a natural ventilation system is in competition with established mechanical systems that offer precise control over the indoor environment).

The prime advantage that natural ventilation has over mechanical ventilation is, of course, the inherently lower energy consumption arising from the absence of mechanical fans. The prime disadvantage is that, compared to mechanical ventilation, it offers less control over the indoor environment. As a result of this, a major objective of modern natural ventilation design is to improve the level of control offered by natural systems. However, the aim of the control is to ensure that the indoor conditions remain within satisfactory boundaries, rather than the very narrow range achievable with mechanical systems. It is in this area that modern technology (computational and experimental techniques) has enabled more accurate and effective designs.

The design of a naturally ventilated building can be a lengthy process involving several stages, which are likely to be iterative. There is no definitive procedure, but it is believed that dividing the process into four stages is a reasonable description, as shown in Fig. 1.1. The first stage is to assess whether natural ventilation is feasible. Here it is necessary to consider the many factors that can influence the ventilation of the building. The more important factors are briefly described in Sect. 1.2. The second stage is to choose a ventilation *strategy*, i.e. the flow pattern that is required. There may of course be more than one strategy for different times of the day or year. Ventilation strategies are dealt with in Sect. 1.3.

Having chosen the strategy, the next stage is to ensure that the airflow rates through the envelope openings and the required internal air motion can be achieved. It is here that the quantitative part of the design process really begins. It is logical to approach this in two stages. The initial quantitative design stage (stage 3 in Fig. 1.1) is where the envelope openings are designed (size and position). This is covered in Sect. 1.5. The second quantitative stage (stage 4) is where off-design investigations are carried out to gain a better idea of the overall performance of the system (Sect. 1.6). Since the technical difficulties of design arise from the complicated

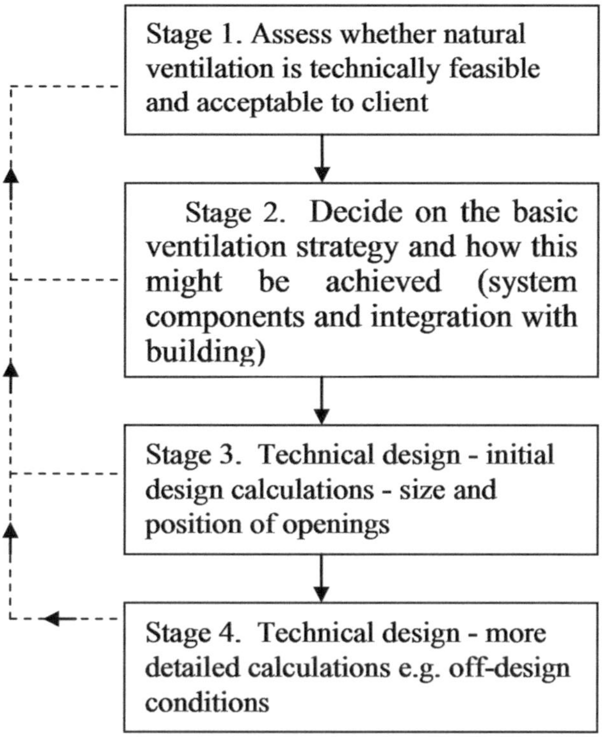

Fig. 1.1 A simplified view of stages in the overall design process

nature of ventilation flows, a brief description of the physical mechanisms is given in Sect. 1.4.

1.2 Feasibility of Natural Ventilation

1.2.1 Climate

Climate is perhaps the most important factor in deciding the feasibility of natural ventilation for non-residential buildings, partly due to variability (Fig. 1.2). The biggest challenge probably lies in hot and humid climates.

The prevention of overheating of occupants during periods of high external temperatures is a major challenge. There are basically two approaches to solving this problem. The first is the use of high ventilation rates, with the aims of (a) convective and evaporative cooling of occupants by high internal air speeds and (b) the removal of internal heat gains by the ventilation air. With high wind speeds, the aim is essentially to achieve an internal temperature that is equal to the external temperature. However high wind speeds are not guaranteed and with low wind speeds the generation of high ventilation rates requires that the internal

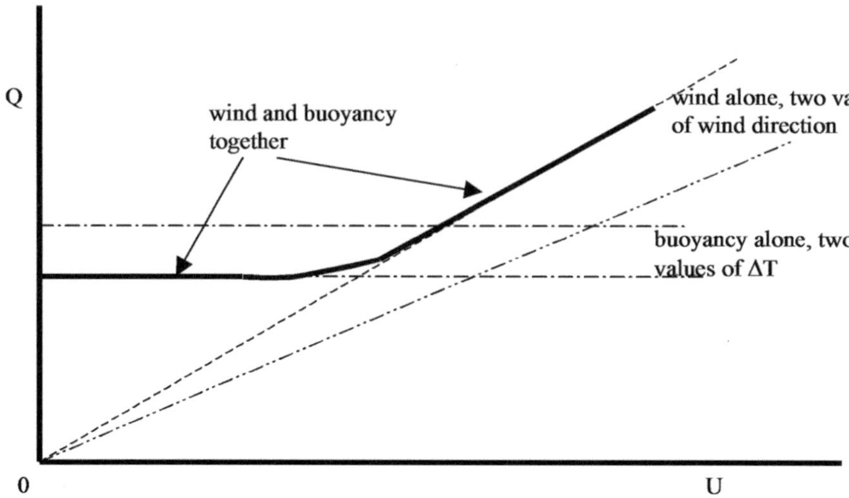

Fig. 1.2 Variability of natural ventilation rates due to climate (for fixed opening size)

temperature be higher than the external temperature by, say, 3°. This may not be acceptable when external temperatures frequently exceed 25 °C. Furthermore, large openings may not always be acceptable in commercial buildings.

The second approach is the strategy of *night cooling*. Here the aim is to cool the internal fabric of the building at night so that it can absorb internal heat gains during the day. This approach requires high ventilation rates at night and low ventilation rates during the day. A successful night cooling design can maintain the internal temperature below the external temperature (see Sect. 1.2.7) during the day, but it requires a relatively high level of control over the ventilation rates.

Perhaps the major weakness of natural ventilation is that on its own it can do little to reduce humidity levels, apart from removing moisture generated internally, e.g. by evaporative cooling of the occupants. There is no direct equivalent to night cooling. This does not mean that natural ventilation cannot be used in humid climates. It clearly can, as witnessed by domestic dwellings and by some commercial buildings (Salmon 1999).

1.2.2 Occupants

Clearly the internal environment produced by the system must be acceptable to the occupants. In fact, the success of a natural ventilation design relies on the behaviour of the occupants for several reasons. First, the occupants are assumed to modify the sizes of openings and to take other actions (e.g. use desk fans to promote cooling; reduce internal gains) in order to obtain acceptable conditions. Surveys indicate that occupants generally favour having control over their environment in this way.

Second, achievement of thermal comfort relies on the willingness of occupants to adapt actively to a changing internal environment by varying their dress and possibly their activities to achieve comfortable conditions. Furthermore, passive adaptation is important. It is increasingly being recognised that comfort criteria for air-conditioned buildings are not appropriate to naturally ventilated buildings. For example, humans adapt to their surroundings in such a way that higher internal temperatures become acceptable when the external temperature increases (Brager and de Dear 2000).

1.2.3 Building Shape and Environment

Both the shape of the building and the environment in which it is situated determine the pressures generated by the wind on the surfaces of the building and the local flow of air around the building. To some extent, the shape of the building can be specifically designed to enhance these effects (Allard 1998).

1.2.4 Building Plan and Layout

Deep plan buildings present problems for natural ventilation, because the entry of fresh air into spaces is concentrated at the external envelope. However a deep plan building can be converted into narrow plan by means of an internal courtyard or atrium, and this is a fairly common strategy to provide crossflow ventilation. High-rise buildings can be naturally ventilated by isolation (in terms of airflows) of the individual floors. There are many high-rise residential buildings that are naturally ventilated in this way. When there is a significant connection (in terms of airflow) between the various floors, the building needs to be treated as a single space. The feasibility of naturally ventilation for very tall buildings (skyscrapers) has certainly received attention (Yeang 1996; Daniels et al. 1993), but currently there are few skyscrapers that rely purely on natural ventilation.

1.2.5 Building Envelope

The building envelope is one of the most important factors in natural ventilation design. In addition to containing the openings, it can influence such factors as thermal storage (for night cooling) and internal heat gains arising from solar radiation.

A major design aim is to specify the size and position of purpose-provided openings in the envelope. Envelopes also contain adventitious openings (i.e. openings that are not purpose provided). As a general rule, the aim should be to minimise adventitious openings, and this is now recognised in many countries by the adoption of standards for adventitious leakage. In the past, much attention has been paid to the measurement and modelling of adventitious leakage (Etheridge and Sandberg 1996). One intention of standards is to eliminate the need to account for adventitious leakage. Adventitious openings can certainly be ignored when the purpose-provided openings are large, e.g. summer conditions, but they may be significant under other conditions.

1.2.6 Internal Heat Gains

Ventilation removes internal heat gains (from lighting, solar radiation, office equipment, etc.) but only at the expense of increased internal air temperature, which in summer conditions is undesirable. This means that minimising internal heat gains during the cooling season is a crucial part of a natural ventilation design. As a rough guide, the internal heat gains should be less than 30 W per m^2 of floor area. Larger values may require some form of additional cooling (passive or active).

1.2.7 Thermal Storage (Night Cooling)

The ability of the envelope of the building to store heat can have a significant influence on the temperatures experienced in a space. The heat transfer rates associated with temperature differences between the air and the fabric can be of similar magnitude to the heat transported by the ventilation air. The night cooling (night ventilation) strategy takes advantage of the thermal storage offered by the fabric of the envelope, i.e. the use of high ventilation rates at night, with the aim of removing heat from the fabric that has been gained/stored during the day. During the following day, the fabric can then absorb heat from the internal air, thereby providing passive cooling. For this strategy to be effective, it usually requires all or most of the following conditions to be satisfied:

- A large temperature swing from day to night ($>$10 K)
- Low internal heat gains ($<$30 W/m^2 floor area)
- Heavyweight construction (high thermal mass)
- High heat transfer coefficient to the internal surface
- High ventilation rates at night, low during the day

For this reason, theoretical predictions of the internal temperatures in a naturally ventilated building should include the dynamic thermal performance of the envelope. Figure 1.3a, b illustrate some of the above points using calculated results for a single office (20 m^2 floor area) with an average internal heat gain from 09.00 to 17.00 of 20 W/m^2 (i.e. a total daily gain of 3.2 kWh). The thermal mass of the surfaces has been assumed to reside in a single wall, for which the one-dimensional

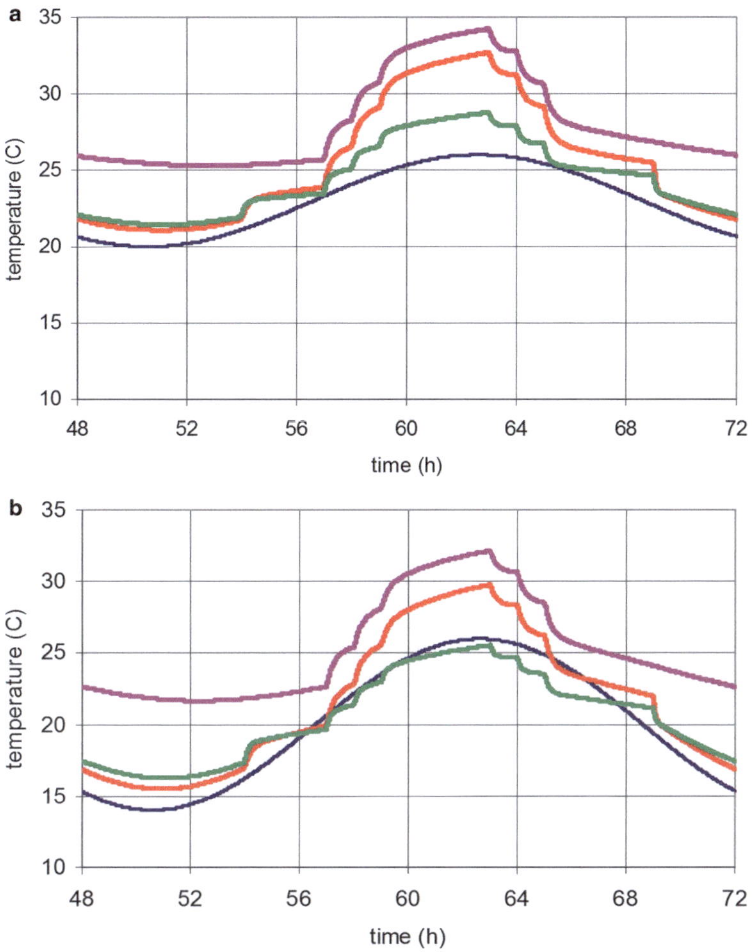

Fig. 1.3 Theoretical results for a simple building envelope, illustrating the potential for night cooling. (**a**) External temperature swing of 6 °C, (**b**) temperature swing of 12 °C

Fourier heat transfer equation has been solved. Each figure shows four curves. The lowest curve is the external temperature variation, which is sinusoidal with a peak value of 26 °C. The uppermost curve shows the internal air temperature when the ventilation rate is 1 h^{-1} at all times. The next curve shows the effect of increasing the ventilation rate to 10 h^{-1} at night for night cooling. The curve below shows the effect of doubling the thermal mass of the envelope, again with high night ventilation. Figure 1.3a is for the case of a diurnal temperature swing of 6 °C. Figure 1.3b is for a swing of 12 °C.

Modern buildings can be designed to enhance night cooling. An example is the addition of thermal mass by means of an exposed concrete ceiling to give direct thermal contact with the air. Another is the use of phase change materials to increase the storage capacity of the fabric. For both examples, the heat transfer rate between the air and the fabric is a limiting factor, and semi-active systems have been developed that use fans to increase heat transfer.

1.2.8 Control

One aim of design is to ensure that openings are sized (maximum and minimum values) so that they allow the occupants to exercise sufficient control. It is not only the magnitudes that are important. The points at which fresh air enters and stale air leaves the building are equally important. Thus a second aim of design is to ensure that the required flow pattern (the ventilation strategy) is controlled. Ideally the design should ensure that the envelope flow pattern is fixed for most weather conditions, with the magnitudes of the envelope flow rates (and internal air speeds) to be controlled by the occupants or by other devices. Air vents are available that automatically adjust their area in response to the pressures acting across them. Control can be by some form of BMS (building management system), e.g. adjustment of stack flow rates by varying the position of a damper in response to a measurement of CO_2 concentration. A system should incorporate fixed air vents to ensure that a minimum level of ventilation is always available (in some cases, adventitious openings are assumed to perform this function).

1.3 Ventilation Strategies

Ventilation strategies are described here in terms of the flow pattern that is required. Different strategies may be required at different times of the day or the year. The more common strategies are identified in the following. It is relevant to divide them into two basic categories (isolated and connected spaces).

1.3.1 Isolated Spaces

In some buildings, the spaces or rooms can be considered as isolated (in terms of airflow) from other parts of the building. For this to be true, the openings to other parts of the building must be small in relation to openings in the external envelope. Figure 1.4 illustrates such spaces and possible ventilation strategies. Spaces A and B are examples of single-sided ventilation, with a large single opening and two small openings at different heights. Spaces C and D are examples of crossflow ventilation of an isolated floor, again with large and small openings. In both cases, the flow pattern is that due to the action of wind alone. Space E shows the flow pattern due to buoyancy alone.

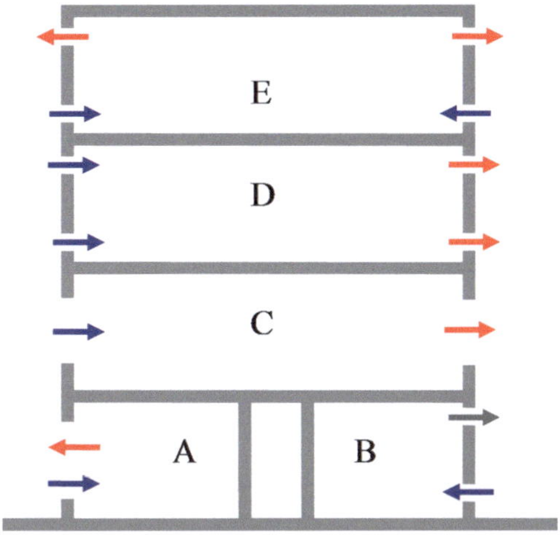

Fig. 1.4 Ventilation patterns for isolated spaces

1.3.2 Connected Spaces: Single-Cell Building

When the spaces in a building are connected by large internal openings, they effectively form a single cell, with the flow through any opening dependent on the flow through the other openings. Such spaces are relatively common in naturally ventilated buildings, partly because of the desire to minimise internal resistance to flow and partly to enhance internal mixing. Figures 1.5 and 1.6 illustrate possible strategies.

In Fig. 1.5a, an atrium is used to generate inward flow of fresh air into all of the occupied floors, i.e. crossflow ventilation of all floors. An advantage of this strategy is that wind and buoyancy can act in unison, provided the outlet opening is in a region of relatively low wind pressure and the internal temperature is higher than the external. The same effect can be obtained by means of an atrium or stairwell at one end of the building (Fig. 1.5b). A "top-down" ventilation strategy can be employed with this arrangement of openings, whereby the arrows are reversed, i.e. fresh air enters through the uppermost opening. Such a strategy may be employed when some form of cooling is provided at high level, e.g. passive downdraught cooling (Francis and Ford 1999). Wind effects are however likely to change the flow pattern.

Figure 1.6a shows the strategy where a chimney or stack is used as the outlet opening to generate fresh air entry. This is similar to Fig. 1.5b, but there is one significant difference. The chimney or stack is in the form of a duct, and the walls of the duct allow the density of air in the stack to be considerably different to that in the space, i.e. the duct walls support a large horizontal variation in density. This allows the forces generated by buoyancy to be increased, e.g. by means of a solar chimney.

Chimneys can also be used specifically to take advantage of wind forces. They are sometimes known as "windcatchers" or wind towers, and there is an interesting

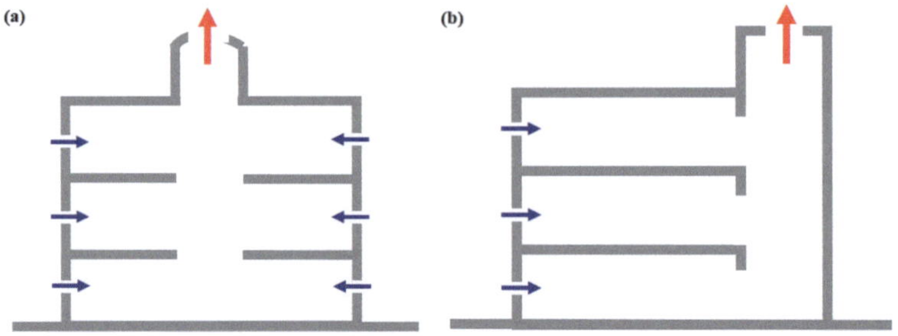

Fig. 1.5 Ventilation patterns based on atria with (a) central atrium, (b) atrium at one end

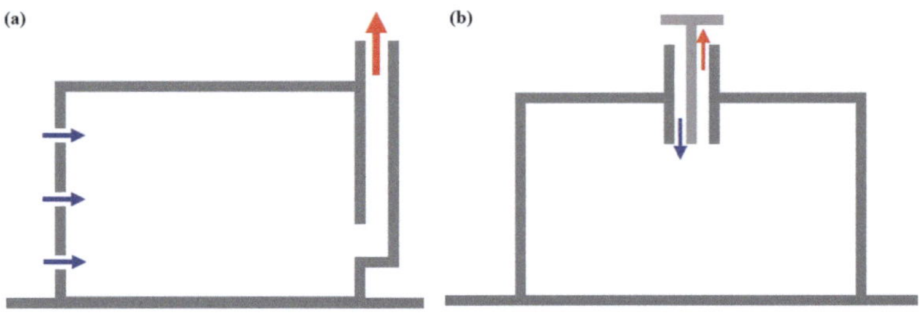

Fig. 1.6 Ventilation strategies based on chimneys (stacks) with (a) one long chimney, (b) two short ducts

variety of traditional designs, particularly in the Middle East (Battle McCarthy Consulting Engineers 1999). Sometimes two separate chimneys are used, with one acting as the inlet and one as the outlet, depending on the wind direction and provided there are no other openings. Sometimes the chimneys are placed together, as illustrated in Fig. 1.6b. The specific type shown in Fig. 1.6b has relatively short chimneys, and because of its short length, it does not protrude far into the space (the buoyancy-generated flow is however relatively low). It can be described as a balanced ventilator, since the intention is that the inlet and outlet flows are equal and this will apply for any wind direction.

1.3.3 Connected Spaces: Multicell Building

When the openings between spaces are neither large nor small compared to the openings in the external envelope, the building can be described as multicell. This description arises from envelope flow modelling. In terms of envelope flow pattern, there may be no difference to the single-cell equivalent.

1.4 Mechanisms: Physical Descriptions

The ventilation of a building can be considered as comprising two processes:

- Envelope flow – the entry and exit of air through openings in the building envelope
- Internal air movement – the motion of the air whilst it is inside the envelope

Internal air motion is of importance, primarily because of its direct influence on the well-being of the occupants (thermal comfort, air quality). With natural ventilation, the internal air movement can have a direct effect on the envelope flows, and by virtue of the fact, it plays a role in determining the temperature (density) distribution within the space and hence the pressure differences across the openings. Thus the two processes are not independent, but the dependence is not always very strong, and in design, it is not uncommon to treat them separately. This is justifiable in the early stages of design, because of the considerable simplifications it offers. The following physical descriptions are for three different conditions, namely, buoyancy alone, wind alone and buoyancy and wind combined. Attention is focused on the difficulties that arise with theoretical modelling and scale modelling.

1.4.1 Buoyancy Alone (No Wind)

In the absence of wind (still-air conditions), the envelope flows are generated by buoyancy, i.e. gravity acting on density differences. If we consider the building shown in Fig. 1.5a, for example, the external air pressure variation is given by the hydrostatic equation, i.e. a linear decrease of pressure P_E (Pa) with height z (m).

When the air inside the building is not in motion, the hydrostatic equation again applies, so the internal surface pressure, P_I, decreases linearly with z (when the internal density is uniform) but at a different rate. At a given height, it can be seen that P_I and P_E are not in general equal, i.e. a pressure difference will be generated across the envelope, and where there is an opening, a flow occurs.

Inside the space, there will be air motion, arising partly from the momentum of the air flowing through the openings and partly due to buoyancy forces. The latter arise primarily from heat transfer into and out of the air at surfaces. Common sources of heat transfer are conduction from the surfaces of the envelope, from heating or cooling appliances, from lighting and from the occupants. These can lead to identifiable convective streams known as plumes. The flow of air in the plumes is usually turbulent. Dry air is transparent to thermal radiation, but radiation still plays a role in the sense that it affects the surface temperatures of the envelope. The fact that surface heat transfer plays a major role in determining the internal air motion leads to major problems in both theoretical and experimental modelling, as discussed in Sect. 1.6.3 and Etheridge (2007b).

Variations of internal density with height (stratification) can be expected, particularly in tall spaces, such as atria, where the temperature at high level is often greater than the temperature at low level. A decrease of density with height often leads to stable stratification, whereas the introduction of dense air at high levels can lead to instabilities, typified by low-frequency unsteadiness. In some cases, negative buoyancy can occur, e.g. when night cooling is employed. In the so-called top-down ventilation system, pre-cooled (and denser) air at high level is induced to flow downwards, thereby providing cooler air at low levels.

The internal velocities generated by buoyancy are often small compared to the velocities of the air through openings, and when this is the case, the internal pressure can be taken as given by the hydrostatic equation. This assumption is made in envelope flow models. A related assumption is that density gradients in the horizontal directions are small, i.e. density varies only in the vertical direction.

1.4.2 Wind Alone (Uniform Density)

With wind alone, the pressure differences across the envelope arise as a result of the changes in velocity that occur as the airflows around the building. The airflow is invariably turbulent, due partly to the turbulence in the upstream flow and partly to flow separations on the building itself.

The fact that the external pressure field is fluctuating means that the flows through the openings will be unsteady, whereas, with buoyancy alone, they are inherently steady. In addition, the velocity conditions at the inlet (or outlet) of an opening can influence the flow characteristic of the opening. A further complication arises when an opening is large in relation to surface in which it lies, when it can affect the flow over the external surface and the external pressure distribution. These effects are difficult to model theoretically, but they present less of a challenge to scale modelling. Small openings are easier to model theoretically and a relatively simple type of model (envelope flow model; see Etheridge (2007b)) can be used. With such openings, the external airflow can be considered as independent of the openings. Furthermore, the internal velocities generated by flow through the openings are small, and the internal conditions are therefore close to those in a laboratory test where the discharge coefficient of the opening can be determined. Both of these features mean that each opening can be treated in the model as a separate entity.

With large openings, the internal motion will be turbulent (characterised by nominally random fluctuations in velocity and pressure throughout the space) with similar velocities to those through the opening. The underlying problem in modelling wind-alone conditions is the presence of turbulence (external and internal). The absence of buoyancy makes scale modelling less difficult, and environmental wind tunnel techniques have been developed that allow useful results to be obtained (see Etheridge (2007a)). Theoretical modelling is also less difficult and the modelling of turbulence is becoming increasingly more realistic.

1.4.3 Wind and Buoyancy Combined

In the general case, buoyancy and wind act simultaneously, and when the two effects are of similar magnitude, the difficulties of theoretical and scale modelling are at their most severe.

1.4.4 Flow Through Openings

1.4.4.1 Types of Opening

The flow characteristic of an opening is determined partly by its geometry (shape and size) and partly by the conditions surrounding the inlet and the outlet. Openings in envelopes fall into two categories, i.e. purpose-provided and adventitious openings. Purpose-provided openings are intentional openings, such as vents, open windows and stacks. A major objective of design is to determine the size (maximum and minimum values) and position of these openings. Adventitious openings are unintentional openings, such as gaps and cracks in walls, doors and windows. Adventitious openings are simply all those openings that are not purpose provided. It is probably impossible to construct a building envelope that is completely airtight. However, excessive adventitious leakage is undesirable if the system is to operate satisfactorily in winter conditions. Under these conditions, the purpose-provided openings will be small, to minimise heat loss, so adventitious openings may be significant. Adventitious openings can be taken account of in the design, although their effect may be the source of considerable uncertainty. With the larger openings associated with summer (cooling) conditions, adventitious leakage can be neglected.

For the purpose of this lecture and its companions, it is convenient to divide purpose-provided openings into the three types shown in Table 1.1. Adventitious openings are the fourth type.

Type 1, 2 and 3 openings are distinguished by their size and shape and the consequential flow characteristics. In Table 1.1, L/d denotes the length-to-diameter ratio and A_{op}/A_w denotes the ratio of opening area to the area of the wall in which it lies. Type 1 openings have low values for both these ratios (NB; the values quoted are only a very rough guide). The low L/d means that the discharge characteristic, C_d, of the opening can usually be treated as a constant. The low area ratio means that the presence of the opening only has a local effect on the pressure distribution generated by the wind on the wall. It also means that flow through the opening will be unidirectional. These features make such openings particularly amenable to treatment in theoretical envelope flow models.

Type 2 openings are long openings such as chimneys and stacks, with L/d typically greater than 5. This means that the discharge coefficient will vary with flow rate. However, it also means that the flow will be unidirectional, so treatment

Table 1.1 Types of envelope openings

Type	Description	Area ratio	L/d	Characteristics	Flow type	Examples
1	Purpose-provided opening: small and short	<0.1	<2	C_d = constant	Unidirectional flow	Vents, small window
2	Purpose-provided opening: long	<0.1	>5	C_d depends on Re	Unidirectional flow	Chimneys
3	Purpose-provided opening: large and short	>0.2	<2	C_d = constant with unidirectional flow	Bidirectional flow possible	Large open windows
4	Adventitious	–	–	C_d depends on Re	Unidirectional flow	Cracks in door and window frames

in an envelope flow model is not too difficult. Type 3 openings, such as large window openings, present the greatest difficulty, due to the large area ratio. This means that bidirectional flow can occur in the presence of wind and buoyancy. Furthermore, they may be large enough to influence the surface pressure distribution and may interact strongly with the local turbulence velocity field. Although the discharge coefficient is likely to be independent of flow rate, when determined in a laboratory, the use of this C_d to describe the flow characteristic becomes questionable.

The above comments relate to theoretical modelling of the openings. When scale modelling is concerned, the key characteristic is whether C_d is dependent on flow rate, so that type 2 openings present the major difficulty (see Etheridge (2007a)). Adventitious (type 4) openings are often unidentifiable and they often have small dimensions. They can be dealt with theoretically but only by making some broad assumptions. It is virtually impossible to include them in scale modelling.

1.4.4.2 Flow Characteristics

The steady flow characteristics of purpose-provided openings with unidirectional flow can be measured in a laboratory test rig (under still-air conditions) by subjecting the opening to a known and steady pressure difference and measuring the resulting flow rate. The flow characteristic can then be conveniently expressed in terms of a discharge coefficient C_d:

$$C_d \equiv \frac{q}{A}\sqrt{\frac{\rho}{2\Delta p}} \tag{1.1}$$

where Δp denotes a defined pressure difference across the opening (Pa), ρ the density of the inflowing air (kg/m^3), A a defined area(m^2) of the opening and q the volume flow rate (m^3/s). In general, C_d will vary with flow rate (Reynolds number

Re). However, for type 1 and type 3 openings, where the flow is dominated by flow separation at sharp edges, the discharge coefficient will be found to be virtually constant above a certain flow rate. This fact is of special importance to scale modelling (Etheridge 2007a), and it also simplifies theoretical calculations.

When the opening is situated in a turbulent flow field (fluctuations of pressure and local velocity), the concept of the discharge coefficient becomes increasingly less tenable as the time-averaged pressure difference across the opening approaches zero. An extreme example of this occurs when the mean pressure difference across the opening is equal to zero, as can arise with a single opening in an otherwise sealed room. Here the ventilation rate becomes independent of the area of the opening, since inward and outward flow is determined by the compressibility of the air in the space. When there is more than one opening, turbulent diffusion by the velocities becomes important. Early work on these and related issues is discussed in Sect. 3.2 of Etheridge and Sandberg (1996).

1.4.5 Mathematical Models

The fundamental equations that govern air motion are the unsteady Navier-Stokes equations (obtained by applying Newton's second law to the motion of a small element surrounding a point in the flow), the corresponding equations for conservation of energy (kinetic and thermal) and mass and the thermodynamic equations of state. Numerical techniques for solving certain forms of these equations have been developed and are generically known as computational fluid dynamics (CFD) (see Sect. 1.6.3). Steady flow is commonly assumed, although unsteady flows can be dealt with.

Another type of model that is commonly used in ventilation design is the envelope flow model. This type is much simpler. It calculates the flow rates through the openings using equations derived from integration of the fundamental equations over large volumes. The internal density (temperature) distribution has to be specified, along with the external wind pressure distribution and the flow characteristics of the openings. Thus the simplicity of the equations is accompanied by a need for empirical data, which may be difficult to obtain. This type is discussed briefly in Sect. 1.6.1 and in detail in Etheridge (2007b). Again, steady flow is commonly assumed, but unsteady models have been developed.

The other type of model in common use is that which combines a model of the thermal behaviour of the envelope with an airflow model. This type is briefly discussed in Sect. 1.6.4. These models by their very nature are unsteady models, but the timescales involved are relatively large (of order 1 h), since the intention is to model the thermal behaviour of the envelope, rather than the effects of wind turbulence.

1.5 Initial Design Process: Size and Position of Openings

Here we are concerned with stage 3 of the design process (see Fig. 1.1). For the chosen strategy (flow pattern), the initial design aim is to determine the sizes (and positions) of openings required to give the required flow rates (magnitudes). This can be done using an envelope flow model in what is known as the explicit method, and the procedure is described in detail in Etheridge (2007b). The starting point is, where relevant, to divide the spaces in the building into two types, i.e. spaces which are isolated and spaces which are connected. As noted earlier, a room or space can be considered as isolated when the openings connecting it to the remainder of the building are very small in relation to the openings in its external envelope. Figure 1.4 shows some examples. Spaces with relatively large openings between them can be considered as connected, in the sense that there is negligible pressure difference between them. In these cases, it is necessary to consider all of the spaces simultaneously when calculating ventilation rates. If the building contains both isolated and connected spaces, they can be treated separately. If the openings connecting the spaces are of similar size to the other envelope openings, it is possible to make use of a multicell envelope flow model. However this cannot be solved in an explicit manner.

1.5.1 Design Conditions

In a climate where there are two distinct seasons that require respectively heating and cooling, two basic design conditions exist. One determines the minimum sizes of the openings and the other determines the maximum sizes. By determining the maximum and minimum openings, the occupants (or other control systems) should be able to exercise the control to satisfy most weather conditions throughout the year. There may of course be other design conditions for which the explicit method is applied; e.g. with night cooling large openings may be required at night and small openings during the day. However, here we consider the two basic conditions.

Ventilation is required to provide satisfactory indoor air quality and it plays a role in satisfying thermal comfort. In an office, the required fresh air rate for air quality will lie in the range from 5 to 8 l/s per person, corresponding, respectively, to control of CO_2 and body odours. From this, it is relatively easy to specify the required envelope flow rates. It is more difficult to specify flow rates required for thermal comfort, partly because the internal temperatures will depend on the dynamic thermal behaviour of the building and partly because the occupants can be assumed to be adaptable.

1.5.1.1 Winter (Heating Season)

A common aim of the winter design condition is to ensure adequate indoor air quality under most conditions, without excessive heat loss due to ventilation. This is not as straightforward as it sounds and there is a degree of choice for the designer. The basic procedure consists of the following steps:

(i) Determine the minimum ventilation rates required for air quality.
(ii) Decide the "worst-case" condition (wind speed, direction and temperatures).
(iii) Use the explicit method to calculate the open areas required for this condition.
(iv) Select appropriate openings.

The areas obtained are basically the minimum sizes for the openings, i.e. the type of openings should be chosen such that this area is permanently available.

The choice of the "worst-case" condition in step (ii) is not clearly defined. The choice should usually correspond to a high ventilation rate; e.g. the designer might choose the weather conditions for which there is only a 1 in 10 chance of them being exceeded. With less extreme conditions, the occupants will be assumed to increase the openings as appropriate. For step (iii), it is preferable to make use of adjustable vents, rather than openable windows. Windows are not ideal for achieving small openings and the coarse control could lead to excessive ventilation.

1.5.1.2 Summer (Cooling Season)

The ventilation rate required to prevent overheating will usually exceed that required for air quality. Thus the first step is to decide the cooling strategy that one wishes to adopt (see Sect. 1.2.1). In climates where the external temperature rarely exceeds 25 °C, the aim may be to achieve high ventilation rates when there is little wind, by allowing the internal temperature to exceed the external temperature by a small amount. For this case, the basic procedure consists of the following steps:

(i) Decide the acceptable peak temperature rise, e.g. 3°.
(ii) Calculate the ventilation rate required to ensure that this temperature rise is not exceeded.
(iii) Take "worst case" to be zero wind speed.
(iv) Calculate the open areas required and design the openable windows such that this maximum value can be achieved.

The magnitude of the acceptable temperature rise in step (i) should lie within agreed comfort criteria. For step (ii), the calculations should ideally include the thermal characteristics of the building, i.e. a full thermal response calculation over a period of several days. Figure 1.3 is an example of such a simulation. In some circumstances, a simple calculation that ignores thermal storage may suffice, i.e. the required ventilation rate is calculated on the assumption that ventilation directly removes all of the internal heat gains.

1.6 Later Design Process

Here we are concerned with stage 4 in Fig. 1.1. This stage is likely to be more demanding in terms of resources than stage 3. For example, once initial sizing has been done, the designer may wish to examine off-design conditions with these openings and to determine more details of the flows and temperatures inside the building. On the basis of these results, the initial sizing calculations may be reconsidered.

There is a wide range of design tools available, which can be summarised under the following headings:

- Scale modelling
- Envelope flow models
- Computational fluid dynamics or CFD
- Combined thermal and airflow models

1.6.1 Scale Modelling

Scale modelling is the subject of Etheridge (2007a). It will be seen that scale modelling is particularly useful for the wind-alone case, primarily because wind tunnel techniques allow relatively accurate modelling of turbulence. Wind tunnel testing is currently the main source of information on wind pressure distributions. Wind tunnels can also be used for measuring directly the ventilation rate of a building and the internal air motion and for investigating the flow characteristics of certain types of opening. These latter uses are however limited by the size of the building model that can be used.

The presence of buoyancy (with wind) imposes severe limitations on scale modelling in wind tunnels, but it is possible to determine ventilation rates. Internal air motion is virtually impossible to deal with. However, scale modelling using liquids rather than air offers some possibilities, particularly for the buoyancy-alone case.

1.6.2 Envelope Flow Models

Envelope flow models are the subject of Etheridge (2007b). As noted earlier, they can be solved in an explicit manner, where the flow rates are specified for given weather conditions and the areas are calculated. More generally, they can also be solved in an implicit manner, where the opening areas are specified and the flow rates are calculated.

Implicit solutions are useful for looking at off-design conditions. Multicell envelope flow models are the general case of single-cell envelope flow models.

They include the effect of the internal partitions, not only in the sense that partitions can increase the resistance to flow through the building but also in the sense that they allow temperature differences between rooms to be defined. The equations that are solved and the assumptions that are made are in most respects the same as for single-cell models. Some multicell models include equations for predicting pollution concentrations. Multicell models can also form part of a combined thermal and flow model. Zonal models are a step further than multicell models, in that the rooms are further divided into a number of zones. Envelope flow models have the advantage that they can be used for assessing the influence of adventitious leakage.

A disadvantage of envelope flow models is that they ignore any interaction between the internal flow and the envelope flow; e.g. the internal density distribution usually has to be specified. It is also difficult to deal with large openings, and the discharge coefficients of the openings have to be specified. The main advantage lies in their simplicity and speed of use. It is also possible to deal with unsteady wind effects, although this requires knowledge of the unsteady wind pressures.

1.6.3 Computational Fluid Dynamics (CFD)

CFD is perhaps the most detailed and potentially the most versatile of all design procedures for ventilation. It can be extended by adding further equations and empiricisms to predict such things as pollutant concentrations, ventilation efficiency and the spread of fires. Here we are only concerned with an overview of CFD and its limitations compared to other procedures.

The term CFD refers to numerical solution of the partial differential equations governing a flow field, such that the velocities, temperatures and pressures at all points in the field are predicted (Etheridge and Sandberg 1996; Blazek 2001; Awbi 1991). For the wind-alone case, the governing equations are the unsteady Navier-Stokes equations (momentum) and conservation of mass. Most CFD models use a time-averaged form of the momentum equation, with semiempirical equations for the turbulent shear stresses. These are known as RANS models. Direct numerical solution (DNS) of the equations is possible, but it is not currently feasible for design purposes. LES (large eddy simulation) lies somewhere between RANS and DNS models and is being increasingly applied to ventilation problems. In the LES model, the larger turbulent eddies are directly modelled, and as a consequence, the turbulent fluctuations are predicted. The inclusion of buoyancy requires simultaneous solution of the thermal energy equation and equations of state. Most CFD models have this capability.

The applications of CFD of prime interest to the design of naturally ventilated buildings include:

(a) Internal flows – calculation of velocity and temperature fields in spaces with specified boundary conditions at the envelope

(b) External flows – calculation of airflow around buildings and surface wind pressure distributions
(c) Whole-field calculations, i.e. the combination of (a) and (b)
(d) Component flows – calculation of flow through components, e.g. discharge coefficients of chimneys

Of these, (a) is currently the most used in design, because it provides a means of determining internal air motion and temperatures which are very difficult to obtain in other ways. However it does require specification of difficult boundary conditions. Applications (b), (c) and (d) are used but not to the same extent.

1.6.3.1 CFD Software

Commercial CFD software is widely available. Most software packages consist of three distinct parts:

(i) The preprocessor, where the user sets the boundary conditions for the solution, including the specification of the computation grid and the number of cells. The number of cells is likely to be of order 10,000 for simple cases such as a small room and of order 1,000,000 for whole-field calculations. For the latter type of calculation, it is necessary to model the upstream and downstream flows, the flows through the openings and the internal flow. This requires a high concentration of cells in and around openings. Figure 1.7 from Yang (2004) indicates the large number of cells required for the relatively simple case of an isolated building with two openings.
(ii) The solver, which carries out the numerical solution. This is an iterative procedure, which continues until convergence is obtained, i.e. a stable solution is reached with error terms below predefined limits. Solution times can vary from a few hours to several days, depending on the complexity of the problem. The user has to assess whether convergence has been satisfactorily achieved.

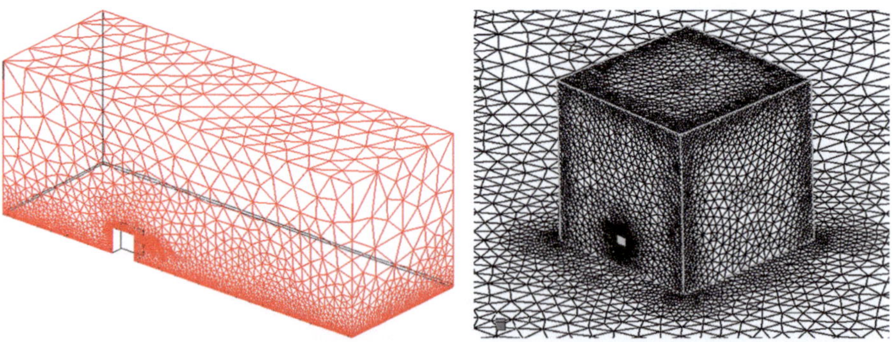

Fig. 1.7 Example of cell distribution for a whole-field calculation of a simple building with two openings (Yang (2004))

This is not necessarily straightforward, partly because the criteria used by the software may not relate to the part of the flow that is of interest and partly because instabilities can occur.

(iii) The post-processor, where the results are presented in a suitable form.

1.6.3.2 Uncertainties in CFD

CFD is a powerful design tool, in the sense that very complicated flow situations can be modelled. However it is important to appreciate that the solutions obtained are subject to uncertainties as with any modelling procedure. Three sources of uncertainty are briefly discussed in the following:

(a) Calculation grid. The results should be independent of the chosen grid and the number of cells. This source of uncertainty should eventually be largely eliminated by the development of more powerful computers.

(b) Assumptions inherent in the equations. Although CFD solves fundamental equations, it still relies on assumptions and approximations. The main ones relate to the properties of turbulence and to heat transfer, particularly close to surfaces. Commercial CFD codes usually offer a range of turbulence models, and the user can investigate the effect of changing the model.

Figure 1.8 (Yang 2004) is an example of the prediction of wind pressures on a simple cuboid building. Full-scale measurements are shown by the solid points. The precise details are not relevant here, but the figure illustrates the uncertainties that

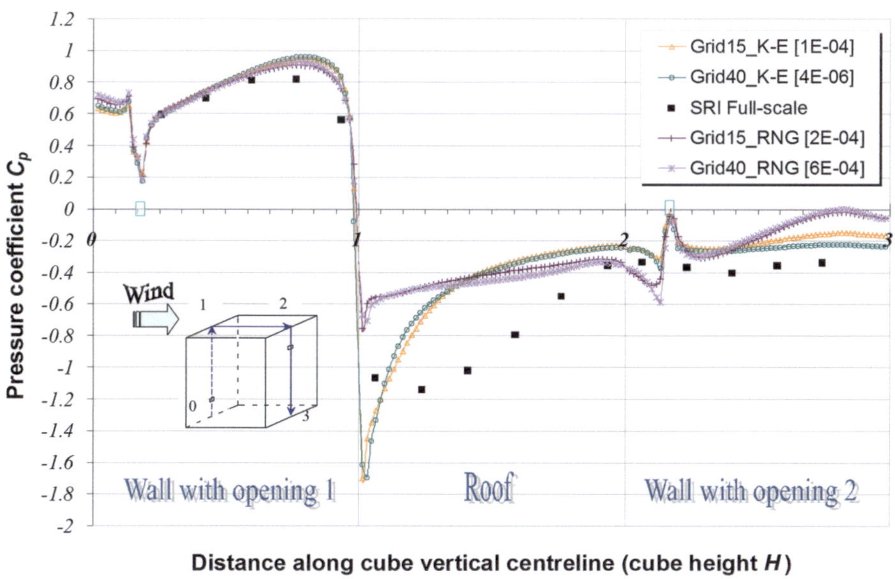

Fig. 1.8 CFD predictions of surface pressure distributions with two turbulence models (From Yang (2004))

can occur with different turbulence models. Turbulence models are still the subject of considerable research and development, and one can expect the uncertainties associated with them to reduce.

(c) Boundary conditions. A real flow satisfies certain conditions at its boundaries and develops with time from an initial condition. However, the majority of CFD calculations ignore the development with time and predict a steady-state solution for given steady boundary conditions. There are at least two sources of uncertainty with this steady approach. First, it is assumed that the steady state is independent of the initial boundary conditions. This is a questionable assumption, particularly when buoyancy is involved. Even with envelope flow models, multiple solutions can occur. Second, there will be uncertainties in the specification of the steady boundary conditions. The first source can in principle be overcome by carrying out an unsteady calculation from an initial boundary condition, but this is a lengthy process. There also remains the fundamental issue that the manner in which the numerical procedure reaches a converged solution is iterative, and this is not the same as the physical process. In some cases, a converged solution may not be obtained. These problems seem to be more prevalent when buoyancy is involved, and this may reflect instabilities in the motion that physically occur. More information on such problems can be found in the Chartered Institution of Building Services Engineers (2005), Etheridge and Sandberg (1996) and Orme (1999).

The value of detailed CFD solutions can be undermined if the boundary conditions are uncertain and the detail is sensitive to the conditions. Steady boundary conditions for velocity are easy to specify in some respects, but thermal boundary conditions at surfaces are more difficult. Unsteady boundary conditions, such as are required with LES models, are difficult to specify. In the long term, when numerical methods and turbulence models have further advanced, the uncertainties of CFD solutions may be determined mainly by the accuracy with which the boundary conditions can be specified.

1.6.4 Combined Thermal and Airflow Models

One of the difficulties in designing natural ventilation systems is the estimation of internal temperature distribution. Especially in summer conditions, the temperature of each space will depend on the ventilation rate, which will itself depend on the temperature distribution, particularly when using buoyancy-driven strategies.

By combining a ventilation model (envelope flow or CFD) with a thermal model, this difficulty can in principle be overcome, although it still represents a challenge. Ideally the two models would be completely integrated, such that the governing equations are solved simultaneously. A simpler approach is to solve the two models separately, with some form of link between their solutions. Examples of such approaches and the problems encountered can be found in CIBSE (2005) and Allard (1998).

1.7 Conclusions

The overall design process can be described in terms of four stages. In the first stage, the feasibility of natural ventilation is assessed. The greatest challenge is likely to be the cooling season. Here it is necessary to reduce internal heat gains. The thermal characteristics of the envelope are also important, as are the ability and willingness of occupants to adapt to relatively high internal temperatures. Night cooling can play an important role.

In the second stage, the ventilation strategy (or strategies) are decided. The choice of strategy (flow pattern) will depend on the purpose of the building and its layout.

The aim of the third stage is to determine the size and positions of the openings that will give the required flow pattern (strategy) and the required envelope flow rates under specified conditions. The maximum and minimum sizes of openings that will allow the occupants to exercise control over the system are determined. Envelope flow models, solved in an explicit manner, are suitable for this purpose.

The fourth and final stage consists of more detailed and wider examination of the design, based on the results of the third stage. The greatest challenge lies in the determination of internal air motion and temperatures in the presence of both wind and buoyancy. Potentially the most versatile design tool at this stage is CFD. However, the use of CFD for certain conditions (e.g. whole-field calculations) is far from routine and is still subject to considerable difficulties and uncertainties. Ultimately the performance of CFD is likely to be limited by the uncertainties present in the specification of boundary conditions. Envelope flow models have a role to play in this stage, partly because of their ease of use and partly because they can be combined with dynamic thermal simulation models. Scale modelling also has potential, particularly for the wind-alone case, where environmental wind tunnels offer realistic turbulence simulation.

References

Allard F (ed) (1998) Natural ventilation in buildings. James & James (Science Publishers Ltd), London
Awbi HB (1991) Ventilation of buildings. E & F Spon, London
Battle McCarthy Consulting Engineers (1999) Wind towers. Academy Editions/Wiley, Chichester
Blazek J (2001) Computational fluid dynamics: principles and applications. Elsevier, London
Brager GS, de Dear R (2000) A standard for natural ventilation. ASHRAE Journal, October 2000
CIBSE (2005) Natural ventilation in non-domestic buildings, Applications Manual AM10:2005. Chartered Institution of Building Services Engineers, London

Daniels K, Stoll J, Pultz G and Schneider J (1993) The sky-scraper – naturally ventilated ? TopE, European Consulting Engineering Network, Brussels
Etheridge DW (2007a) Scale modelling of natural ventilation. COE International Advanced School on Environmental Wind Engineering, COE-IAS4, Soongsil University, Seoul, December
Etheridge DW (2007b) Theoretical models of envelope flow – steady and unsteady. COE International Advanced School on Wind Effects on Environmental Wind Engineering, COE-IAS4, Soongsil University, Seoul, December
Etheridge DW (2007c) External wind effects on flow through small openings and leakage measurement. COE International Advanced School on Environmental Wind Engineering, COE-IAS4, Soongsil University, Seoul, December
Etheridge DW, Sandberg M (1996) Building ventilation: theory and measurement. Wiley, Chichester
Francis A, Ford B (1999) Recent developments in passive downdraught cooling. In: Lewis O and Goulding J (Eds), European Directory of Sustainable and Energy Efficient Buildings. James & James, London
NATVENT (1997) Overcoming technical barriers to low energy natural ventilation in office type buildings in moderate and cold climates. Building Research Establishment. http://projects.bre.co.uk/natvent
Orme M (1999) Applicable models for air infiltration and ventilation calculations, AIVC TN 51. Air Infiltration and Ventilation Centre, Coventry
Salmon C (1999) Architectural design for tropical regions. Wiley, New York
Yang T (2004) CFD and field testing of a naturally ventilated full-scale building. PhD thesis, University of Nottingham, Nottingham
Yeang K (1996) The skyscraper bioclimatically considered: a design primer. Wiley, Chichester

Chapter 2
Theoretical Models of Envelope Flow: Steady and Unsteady

David Etheridge

Abstract The different types of envelope flow model are first outlined. The basic theory and assumptions for steady envelope models are then described. Attention is then focused on the explicit method of solution used for the initial stage of design. Brief mention is made of the use of envelope models for more general calculations, using the more common implicit method of solution. Examples are given of the sort of calculations that can be done with such models. The theory of unsteady models and possible applications to design are described. The difficulties associated with envelope flow models are outlined.

Keywords Natural ventilation • Theoretical modelling • Building envelope • Unsteady flow

2.1 Introduction

Envelope flows are of considerable importance in natural ventilation design, because a primary design objective is to get the required amount of air into a building under specified conditions. By making certain assumptions about the internal and external flows, envelope flow models enable envelope flow rates to be calculated for specified conditions.

Note by Author This chapter is the unchanged text of the third of four related lectures on natural ventilation, given in 2007 at the COE International Advanced School in Japan and then Korea. Of the three other lectures (Etheridge 2007a, b, c), the first lecture (Etheridge 2007c) is given here as Chap. 1, again in its original form. It should be noted that a more comprehensive and updated treatment of natural ventilation can be found in the recent book by the author (Etheridge DW (2012) Natural ventilation of buildings – theory, measurement and design. Wiley, Chichester).

D. Etheridge (✉)
Old Hunstanton, Norfolk PE36 6EQ, UK

previously Department of Architecture and Built Environment, University of Nottingham, Nottingham, UK
e-mail: etheridge.david@btinternet.com

Y. Tamura, R. Yoshie (eds.), *Advanced Environmental Wind Engineering*,
DOI 10.1007/978-4-431-55912-2_2

2.1.1 Types of Envelope Flow Model

Envelope flow models can be broadly distinguished by whether they are purely empirical or semi-empirical, multicell or single cell and steady or unsteady. This is illustrated in Fig. 2.1. Details of the early development of the various types can be found in Chap. 4 of Etheridge and Sandberg (1996).

All envelope flow models contain some empirical content, i.e. they rely on information gained from experimental measurement. The primary examples of this are the surface wind pressure coefficients and the discharge coefficients of the openings. The origin of the former is usually wind tunnel tests, and the origin of the latter is usually a laboratory test. In addition, most envelope models require the internal density (temperature) to be specified. The simplest form of model is that for a single-cell building, but the basic equations can be applied to multicell buildings. Purely empirical models are rarely used and are not dealt with here (see Sect. 4.9 of Etheridge and Sandberg (1996).

Steady flow models use the truly steady form of the flow equations. Basically they make use of what is known as the *pseudo-steady* flow assumption, i.e. the time-averaged flow characteristic of an opening is the same as it would be if there were no unsteadiness due to turbulence (see Sect. 2.1.6 of Etheridge and Sandberg (1996)). This is valid for buoyancy alone, but it is an approximation in the presence of turbulence generated by the wind.

Unsteady envelope flow models solve an unsteady form of the basic equations, but they too make an assumption concerning the flow characteristics of the openings, namely, the *quasi-steady* assumption, whereby the truly steady flow equation is assumed to be valid at all instants of time. Unlike steady flow models, they require simultaneous records of surface wind pressures. There is a class of envelope model that can be described as a combined airflow and thermal model. The model predicts the internal density distribution by making assumptions about the internal heat transfer mechanisms.

Section 2.2 describes the basic theory behind steady flow models, and Sect. 2.3 describes how they can be used for the initial design by adopting the *explicit*

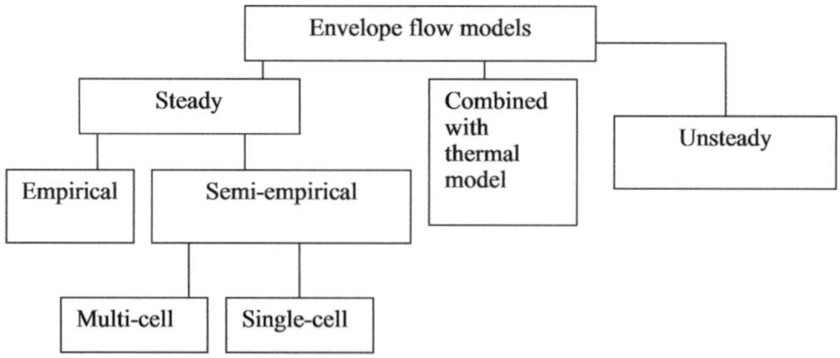

Fig. 2.1 Types of envelope flow models

method of solution. Section 2.4 briefly describes the use of single-cell and multicell models for the later stages in design. The basic theory of unsteady envelope models is described in Sect. 2.5, and some examples of their use for design are given in Sect. 2.6.

2.2 Steady Flow Models

The basic equations for steady flow envelope models can be illustrated by considering an opening in an envelope as shown in Fig. 2.2.

2.2.1 Discharge Coefficient

Of fundamental importance to envelope flow models is the discharge coefficient C_d (−), defined by

$$C_d \equiv \frac{q}{A}\sqrt{\frac{\rho}{2\Delta p}}. \tag{2.1}$$

This provides a relationship between the flow rate q_i (m^3/s) through an opening (subscript i) and a pressure difference Δp_i (magnitude $|\Delta p_i|$ (Pa) and direction S_i (−)) across the opening, by means of the discharge coefficient C_{di} and a specified geometric area A_i (m^2)

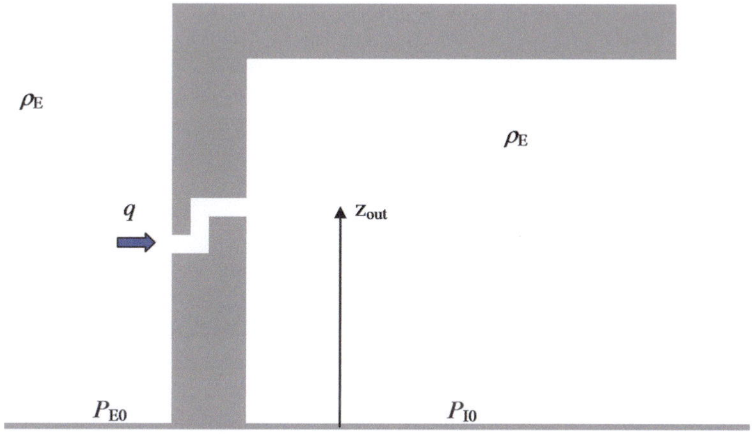

Fig. 2.2 Laboratory measurement of discharge coefficient of an opening

$$q_i = C_{di} A_i S_i \sqrt{\frac{2|\Delta p_i|}{\rho}}. \tag{2.2}$$

The discharge coefficient is defined (and measured) under nominally *still-air* conditions with uniform density, as illustrated in Fig. 2.2. The opening is placed in one wall of a large box or room. The air in the box and the air in the surroundings are at the same temperature, i.e. the density, ρ_E (kg/m^3), is uniform throughout. Since the air is still, the pressures at the inlet and outlet are given by the hydrostatic equation. When the inlet and outlet are at different heights, there will be a difference between the pressures at the two points, but no flow will be generated because the resulting force on the air in the opening is exactly balanced by gravity. To generate the air flow in the laboratory, a fan is used, and this generates a uniform pressure change on the outlet side. The pressure difference Δp is the difference between the pressures at height z_{out} (m) on the two sides of the envelope (or, in fact, at any specified height z).

2.2.2 *Flow Induced by Density Difference*

Now consider the case where the opening lies in a wall of a building and the flow is induced not by a fan, but by a difference in density, as shown in Fig. 2.3. Since the flow is inward, it can be assumed that the density in the opening itself remains equal to ρ_E. It is then clear that up to the outlet, the flow conditions are the same as in Fig. 2.2.

The only difference between the pressure conditions around the outlet is that the change of pressure with z is slightly different on the outlet side. The flow through

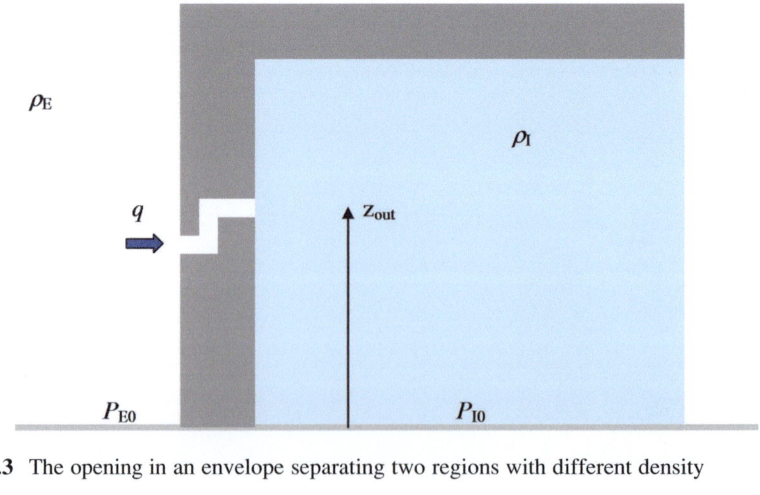

Fig. 2.3 The opening in an envelope separating two regions with different density

the opening will respond to the pressure surrounding the outlet, and it is reasonable to assume that the appropriate pressure is that at height $z = z_{out}$ (away from the influence of the velocities generated by the flow). This pressure is given by the hydrostatic equation as

$$P_{out} = P_{I0} - \rho_I g z_{out}, \tag{2.3}$$

where P_{I0} denotes the pressure (Pa) at $z = 0$. The pressure at the inlet at z_{out} is

$$P_{inket} = P_{E0} - \rho_E g z_{out}, \tag{2.4}$$

where P_{E0} denotes the pressure (Pa) at $z = 0$.

Thus, it follows that

$$\Delta p = P_{E0} - P_{I0} - (\rho_E - \rho_I) g z_{out} \tag{2.5}$$

and hence, with $\Delta\rho_0$ (kg/m^3) denoting $(\rho_E - \rho_I)$

$$q = C_{di} A S \sqrt{\frac{2|P_{E0} - P_{I0} - \Delta\rho_0 \cdot g z_{out}|}{\rho}}. \tag{2.6}$$

When the flow is outward, the same arguments apply, but z_{out} is now the height of the opening on the external surface, reflecting the fact that the density in the opening is now ρ_I. Thus,

$$\Delta p_i = -(P_{E0} - P_{I0} - \Delta\rho_0 g z_{out}). \tag{2.7}$$

The negative sign reflects the fact that the flow is outward, i.e. $S_i = -1$ since q_i is negative.

From the above, it can be seen that the magnitude of the pressure difference in Eq. (2.2) is given by

$$|\Delta p_i| = P_{E0} - P_{I0} - \Delta\rho_0 g z_{out}. \tag{2.8}$$

This equation is quite general and applies to long openings such as chimneys. The important parameter is z_{out}, where z_{out} is the height where the flow *leaves* the opening. This follows from the outlet boundary condition, which states that the pressure at the outlet is determined by the pressure of the surroundings. The height of the inlet to the opening is not relevant, because it is assumed that the density inside the opening is the same as the density of the air at the inlet (the position of the inlet is used to determine the relevant wind pressure; see below). It should be noted that for a long opening aligned in the vertical direction, such as a chimney, z_{out} will change significantly with flow direction.

If the internal density varies with height z, Eq. (2.8) becomes

$$|\Delta p_i| = P_{E0} - P_{I0} - \rho_E g z_{out} + g \int_0^{z_{out}} \rho_I . dz, \qquad (2.9)$$

and this allows non-uniform density distributions to be dealt with.

2.2.3 General Case: Buoyancy and Wind

When the wind blows, the external pressure around the opening has an additional contribution due to the wind, p_{wi}. This simply adds to the hydrostatic pressure equation. Thus, for the general condition,

$$|\Delta p_i| = P_{E0} - P_{I0} - \Delta\rho_0 g z_{out} + p_{wi}. \qquad (2.10)$$

There is an implicit assumption here that the opening is such that it does not change p_{wi}. This is reasonable for small openings such as air vents but less tenable for openings such as windows with opening vanes. An associated assumption is that the external velocity due to the wind does not alter the discharge coefficient. This is a reasonable assumption when the outlet lies in the external flow but is less reasonable when the inlet is exposed to the wind, particularly if the opening has vanes (see Etheridge (2007a)).

It is common practice to obtain p_{wi} from wind tunnel tests where wind pressures are quoted in the form of the pressure coefficient C_{pi} (−)

$$C_{pi} \equiv \frac{p_{wi} - p_{ref}}{0.5\rho U^2}. \qquad (2.11)$$

Thus, the pressure difference across an opening whose inlet or outlet is situated in the external flow at z_i can be written as

$$|\Delta p_i| = P_{E0} - P_{I0} - p_{ref} - \Delta\rho_0 g z_{outi} + 0.5\rho U^2 C_{pi}. \qquad (2.12)$$

2.2.4 Conservation of Mass for the Envelope

The principle of mass conservation applied to the fixed volume defined by the envelope gives a relationship between the flow rates of the openings

$$\sum \rho_i q_i = 0 \qquad (2.13)$$

i.e. the net mass flow into the building is equal to zero. In the context of design procedures, it is an acceptable approximation to ignore the differences between densities, so the equation can be simplified to

$$\sum q_i = 0 \tag{2.14}$$

2.2.5 Solution of the Equations

An envelope flow model thus consists of the following three basic equations:

$$\sum q_i = 0 \quad \text{one equation for the envelope} \tag{2.15}$$

$$q_i = C_{di} A_i S_i \sqrt{\frac{2|\Delta p_i|}{\rho}} \quad \text{one equation for each opening} \tag{2.16}$$

$$|\Delta p_i| = \Delta p_0 - \Delta \rho g z_i + 0.5 \rho U^2 C_{pi} \quad \text{one equation for each opening} \tag{2.17}$$

It is important to include the sign of the pressure difference, S_i in Eq. (2.16). The convention used here is that $S_i = 1$ for flow entering the space and $S_i = -1$ for flow leaving the space. In Eq. (2.17), $P_{E0} - P_{I0}$-p_{ref} has been denoted by Δp_0 for brevity. Although there are three terms in ($P_{E0} - P_{I0}$-p_{ref}), it is only P_{I0} that is a variable; P_{E0} and p_{ref} are constants.

For the general case of N openings, there are $2N + 1$ equations. The unknowns are q_i, Δp_i and Δp_0.

The equations can be solved by determining the value of Δp_0 *at* which the continuity equation is satisfied. In physical terms, when the ventilation pattern of a building is changed from one steady state to another (by opening a window, say), the internal pressure P_{I0} adjusts until the flows through the openings are such that the continuity equation is again satisfied. Mathematically, this adjustment is done by an iterative procedure. When the equations are solved in this way, it is referred to as an *implicit* method.

The equations can be solved directly (i.e. without the need for iteration) by specifying the value of Δp_0 and the values of q_i (magnitude and direction) to find the values of A_i which will give that particular flow pattern. When solved in this way, the model is referred to as an *explicit* method. The explicit method is particularly useful in the initial design stages for sizing openings, such that they give the required flow rates under a specified design condition. Using the calculated areas, the implicit procedure can then be used for off-design calculation of flow rates.

The explicit procedure is best used for openings (types 1 and 3) with a constant discharge coefficient, i.e. C_{di} does not vary with flow rate (Reynolds number). This is the case for most purpose-provided openings. It is not generally true for adventitious openings, for which the implicit procedure is more appropriate, because more assumptions have to be made. Envelope flow models can account for adventitious openings (see Sect. 2.4). Envelope flow models are best suited for dealing with openings with unidirectional flow. For large openings such as windows, it is possible for bidirectional flow to occur, and this is more difficult to treat, although there are ways of doing so (see Sect. 4.11 of Etheridge and Sandberg (1996) for some early approaches).

2.2.6 Main Assumptions in Envelope Flow Models

The main assumptions in envelope flow models are as follows. The air velocities inside the space are assumed to be low, such that the internal pressure is given by hydrostatic equation. This is a reasonable assumption for small openings. For example, using Bernoulli's equation, we can see that an air speed of 0.5 m/s corresponds to a pressure change of about 0.15 N/m^2, which can usually be neglected in relation to the pressure difference across the opening. A further assumption is that the openings are assumed to be small in relation to the size of the walls containing them, such that the external wind pressure distribution is unchanged (apart from a small region around the opening). This is important, since it means that pressures determined in the absence of the openings (e.g. from wind tunnel tests) can be used and that each opening can be treated as a separate entity. It is also assumed that the discharge coefficient of an opening is unaffected by the presence of external flow arising from the wind. This assumption is discussed further in Etheridge (2007a) and Ohba (2007). Similar assumptions are made concerning the internal conditions. In general, there will be air movement inside the space, but with small openings, the velocities are generally low. This will not generally be true when the openings are large in relation to the wall or space containing them. In such cases, the basic approach of treating individual openings separately becomes less tenable. However, the treatment of such flows by envelope models is the subject of current research, e.g. Ohba (2007).

2.3 Initial Design Calculations: Explicit Method

In the following, it is assumed that the openings have discharge coefficients that are independent of flow rate (Reynolds number). This is consistent with the fact that the explicit method is used to determine the sizes of purpose-provided openings. Long openings such as chimneys will be Reynolds number dependent, but C_d will not vary greatly over moderate Re, when the flow rate is close to its design value (this dependency is important for scale modelling; see Etheridge (2007b)).

As noted in Etheridge (2007c), the starting point to the design (in the present context) is to divide the spaces in the building into two types, i.e. spaces which are isolated and spaces which are connected. A room or space can be considered as isolated when the openings connecting it to the remainder of the building are very small in relation to the openings in its external envelope. Two basic situations can be identified, *single-sided* and *crossflow,* as shown in Fig. 2.4.

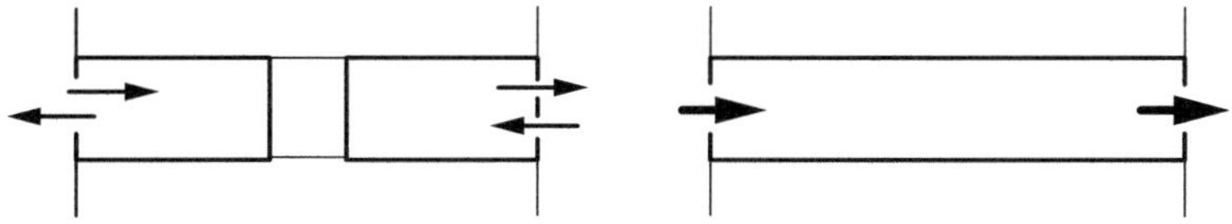

Fig. 2.4 Isolated spaces with single-sided and crossflow ventilation

Simple equations have been developed that allow flow rates to be related to the size of the openings and to the surrounding conditions (e.g. see Etheridge and Sandberg (1996) and CIBSE (2005)). Some of these equations are basically empirical, and research is underway to develop more rigorous relationships, particularly for the crossflow case (Ohba 2007).

Spaces that have relatively large openings between them can be considered as connected in the sense that there is negligible pressure difference between them. This is a fairly common situation, because naturally ventilated buildings often rely on low internal resistance to flow, and it means that they can be treated by single-cell envelope models with the explicit method of solution, as discussed in the following section.

2.3.1 Connected Spaces: Sizing and Positioning of Openings

A typical problem is illustrated in Fig. 2.5 and is taken from CIBSE (2005). The ventilation strategy is that fresh air should enter all of the occupied spaces, with the atrium opening acting as the outlet. The values of q_{i} have been specified for a given condition (temperatures, wind speed and direction). The opening heights z_{i} and the height of the uppermost opening are known. It is required to calculate the areas A_{i} such that fresh air enters each opening at the required rate. The first step in the explicit method is to specify a value for Δp_0. The value has to be compatible with the flow pattern, but there is still some flexibility in the value. Since the flows through the wall openings in Fig. 2.5 are inward, it follows that the pressure difference must change sign at a height which lies somewhere between z_3 and z_4. The height at which $\Delta p_i = 0$ is known as the neutral height z_n. By specifying z_n, one is specifying Δp_0 because it is given by Eq. (2.17) (with wind terms omitted) by putting $\Delta p_i = 0$ and $z = z_n$, i.e.

$$\Delta p_0 = \Delta \rho_0 g z_n \tag{2.18}$$

It then follows that Δp_i is known and is given by Eq. (2.17) as

$$\Delta p_i = \Delta \rho . g z_n - \Delta \rho_0 g z_i . \tag{2.19}$$

The required areas can then be found using a rearrangement of Eq. (2.16)

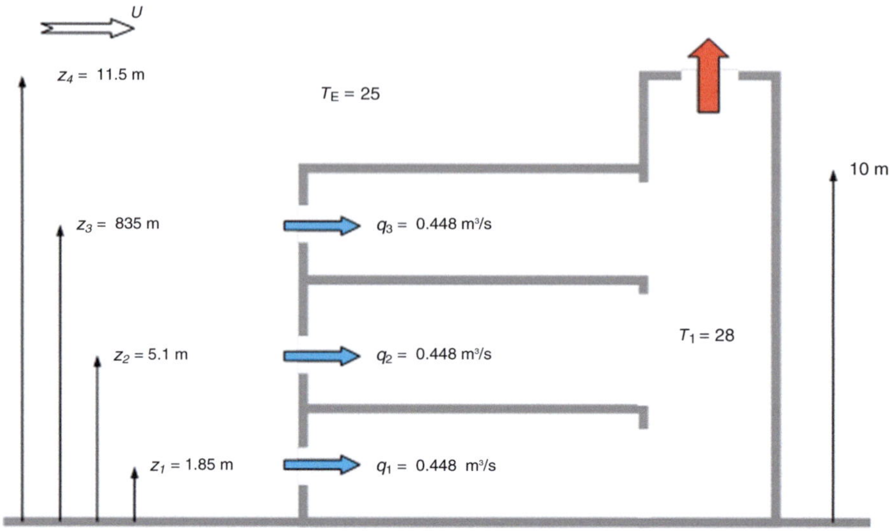

Fig. 2.5 Properties of building for worked example (Reproduced from (CIBSE 2005), CIBSE)

$$C_{di}A_i = \frac{q_i}{S_i}\sqrt{\frac{\rho}{2|\Delta p_i|}}. \tag{2.20}$$

The equations are solved for the product $C_{di}A_i$. There is no need to carry out separate calculations when either C_{di} or A_i is changed, provided other conditions remain the same.

The area of the outlet opening is found in the same way, since its flow rate is given by Eq. (2.15).

2.3.2 Buoyancy Alone, Uniform Internal Temperature

Table 2.1 shows the calculations for a summer design condition with $\Delta T_0 = 3$ (K) and with the neutral height specified as being at the top of the occupied spaces, i.e. 10 m. The value of Δp_0 is first calculated from Eq. (2.17) as 1.185 Pa. Equations (2.17) and (2.20) are then used to determine Δp_i and $C_{di}A_i$.

Although using any value of z_n between 8.35 and 11.5 m would satisfy the flow pattern, it is important to note that a value close to one of these two extremes will lead to a very large area, which is unrealistic. An alternative way of specifying Δp_0 is to set the values of Δp_i for openings 3 and 4 as opposite and equal. This leads to $\Delta p_0 = 1.1762$ Pa, and the resulting areas do not differ greatly from those above.

Specifying Δp_0 completely determines all the Δp_i. It follows that the height of any opening can be changed without affecting the results for other openings. Also the flow rate through any of the *inlet* openings can be changed, without affecting

Table 2.1 Calculated areas for summer design condition

Opening	z_i (m)	C_{pi} (–)	q_i (m^3/s)	Flow pattern	S_i (–)	Δp_i (Pa)	$C_{di}A_i$ (m^2)	C_{di} (–)	A_i (m^2)
1	1.85	0.25	0.448	Inward	1	0.9659	0.3531	0.61	0.579
2	5.1	0.25	0.448	Inward	1	0.5807	0.4554	0.61	0.747
3	8.35	0.25	0.448	Inward	1	0.1955	0.7848	0.61	1.286
4, outlet	11.5	−0.1	−1.34	Outward	−1	−0.1778	2.4692	0.61	4.048

any of the other *inlet* areas. The area of the outlet opening will change, because the outlet flow rate will change when any inlet flow rate changes. In reality a change of an inlet area will lead to changes in the neutral height and the flow rates through all of the openings. This is what is predicted by an implicit method, and the basic difference between the explicit and implicit methods can thus be seen.

With uniform temperature, the flow pattern in Fig. 2.5 is achievable only when the internal temperature is greater than the external temperature. The building may actually be designed to make use of night ventilation with the aim of keeping the internal temperature below the external temperature at all times. Although it is mathematically possible to obtain a solution, it will lead to the sign of Δp_i being opposite to the sign of q_i which is a physical impossibility. However, it is possible to achieve the required flow pattern when the internal temperature is less than the external temperature for cases where the density is not uniform, i.e. where internal stratification occurs and for the case with a chimney (see CIBSE (2005)).

If the user wishes to calculate the case where the internal temperature is less than the external temperature and the flow pattern in Fig. 2.5 is reversed (this can occur as a result of night cooling or downdraught cooling), it is necessary to reverse the signs of q_i and S_i. Thus, if we put $T_I = 25$, $T_E = 28$ and $\Delta T_0 = -3$ in the above example and reverse the signs of q_i and S_i, it will be found that the same areas as those in Table 2.1 are obtained.

2.3.3 *Wind Alone*

Here, the aim might again be to achieve the same flow pattern as in Fig. 2.5. This may be optimistic for some wind directions, because low wind pressures on the leeward side will act to impede fresh air entry through openings on that side. An essential but not sufficient requirement to achieve the pattern is that the wind pressure coefficient on the upper opening must be more negative than the coefficients on the leeward side. For the present simple example, the wind direction is such that the openings are on the windward side with positive C_{pi}. The required flow pattern would be achieved if the internal pressure were such that the pressure difference across the upper opening is one half of the total pressure difference across the building, i.e. $\Delta p_4 = -0.5\rho U^2.\Delta Cp/2$, and this condition can be used to specify Δp_0. The value of Δp_0 is negative, because the flow is outward.

Substituting $\Delta p_4 = -0.5\rho U^2.\Delta Cp/2$ into Eq. (2.17) gives Δp_0, i.e.

$$\Delta p_0 = -0.5\rho U^2\left(\frac{\Delta C_p}{2} + C_{p4}\right), \tag{2.21}$$

which gives the following expression for Δp_i:

$$\Delta p_i = -0.5\rho U^2\left(\frac{\Delta C_p}{2} + C_{p4}\right) + 0.5\rho U^2 C_{pi} \tag{2.22}$$

Equation (2.20) can then be used to evaluate $C_{di}A_i$.

2.3.4 Wind and Buoyancy Combined

The same procedure is employed for the combined case. Combining the weather conditions for examples 1 and 2, the same flow pattern is specified, so the value for Δp_0 is obtained by specifying that the values of Δp_i for openings 3 and 4 are opposite and equal. Applying Eq. (2.17) gives

$$\Delta p_0 - \Delta\rho.gz_3 + 0.5\rho U^2 C_{p3} = -\left(\Delta p_0 - \Delta\rho.gz_4 + 0.5\rho U^2 C_{p4}\right) \tag{2.23}$$

which leads to $\Delta p_0 = 0.771$ Pa.

2.3.5 Non-uniform Temperature

The explicit method can be extended to the case where the internal density is not uniform, by making use of Eq. (2.9). In CIBSE (2005), examples are given where the internal temperature is assumed to have the following variation with z:

$$T + 273 = \frac{T_0 + 273}{\left(1 - \left(1 - \frac{T_0+273}{T_n+273}\right)\left(\frac{z}{z_n}\right)^n\right)} \tag{2.24}$$

where T_0 and T_n denote the temperatures at $z = 0$ and $z = z_n$.

2.3.6 Chimneys and Stacks

Chimneys and stacks can also be treated as described in Sect. 4.3.4 of CIBSE (2005). The temperature in the chimney can be different to that in the space, to simulate a solar chimney. Since C_d does not vary greatly with flow rate for turbulent flow, one can assume C_d is constant.

2.4 Later Design Stage: Implicit Solutions

Envelope flow models are commonly used in the later design stage to investigate off-design conditions and to determine effects that cannot be included in the explicit solution. Multicell models are particularly important here. They have been developed to quite sophisticated forms for this purpose. The early development of such models and some of the problems associated with them are described in Chap. 4 of Etheridge and Sandberg (1996). A well-known example of a current model is COMIS. This has been developed for use in predicting indoor air quality, i.e. CONTAM.

Envelope flow models have also been incorporated in dynamic thermal simulation models of buildings, of which there are several commercial codes available. A much simpler form of a combined airflow and thermal model is that which combines an envelope flow model with the internal flow associated with buoyant plumes (Linden et al. 1990), leading to simultaneous prediction of ventilation rate and temperature stratification within the space. The underlying problem with this approach is the assumptions concerning the manner in which heat is transferred to the air contained within the space. However, the model has been shown to be capable of giving good quantitative agreement with simple physical models. Another interesting outcome from this type of model is that multiple solutions are possible, and this has been confirmed experimentally by tests on simple scale models (Li et al. 2000).

2.4.1 Effect of Adventitious Leakage

One area that is difficult to treat in design is the effect of adventitious leakage. Although the effect can be neglected when the purpose-provided openings are large, it is quite possible that the adventitious leakage of a building will be significant for the winter design case, i.e. it could exceed the leakage associated with the fixed minimum areas. This possibility should be investigated if excessive ventilation heat loss is to be avoided. One approach to the problem is to estimate the adventitious leakage which will give the design outlet flow rate with the same pressure distribution as the air vents (Etheridge 2002). This equivalent leakage can be compared with either the measured value or the range of values likely to be encountered. If it is less than either of these values, there may be no need for a fixed minimum opening. If it is much less, it may be desirable to tighten the building envelope.

When dealing with adventitious leakage, several assumptions have to be made, e.g. the distribution of leakage over the envelope. Of considerable importance is the choice of equation that is used to describe the flow characteristic of the openings (basically to relate the low-pressure leakage to the leakage measured at high pressures). Although the power law is commonly used

$$Q_L = k\Delta p_L^n, \tag{2.25}$$

it is believed that a quadratic equation

$$\Delta p_L = aQ_L^2 + bQ_L \tag{2.26}$$

is preferable. In Eqs. (2.25) and (2.26), k and n and a and b are the leakage coefficients, and Q_L and Δp_L denote, respectively, the leakage and the applied pressure difference. The differences between the two equations are large at low pressures (>40 %), so the choice of equation is important (Chiu and Etheridge 2002).

2.4.2 Data Requirements

The use of envelope flow models requires substantial data input by the user, namely, wind pressures, air temperatures, discharge coefficients and, if possible, the adventitious leakage.

Surface wind pressure coefficients for generic building forms are available, e.g. Orme (1999). For specific buildings, the information can be generated from wind tunnel studies or possibly CFD. It is important that the correct value of wind speed U is used when evaluating p_{wi}, because p_{wi} is proportional to U^2, and, in the absence of buoyancy, flow rates are proportional to U. The C_{pi} data set should specify the height at which the reference wind speed was measured and the nature of the terrain surrounding the building. The designer will probably not have wind speed records for the reference position, and it is then necessary to estimate the values of U from the data available, e.g. from Meteorological Office records for a site that could be many kilometres from the building. A simple procedure is available for doing this, but it is subject to uncertainty associated with changes of terrain between the building and the measurement site. A more recent procedure is that developed for evaluating structural loads on buildings. This would need to be adapted for ventilation calculations, because the aim is to determine the magnitude of a three-second gust, which is not relevant to ventilation. Further details of the procedures can be found in CIBSE (2005). The uncertainty associated with determining wind pressures is one reason why the treatment of ventilation due to wind is less precise than for buoyancy alone. In this connection, it can be noted that for a summer design condition, the wind is likely to be less important than buoyancy, because the main problem is to achieve adequate ventilation under no wind conditions. For a winter design condition, when the temperature difference is large, buoyancy can be the dominant driving force at relatively high wind speeds.

Buoyancy is much easier to treat than wind, because temperatures are subject to much less uncertainty. Accurate records of external temperature are widely available and the appropriate values will have been defined by the weather data used for the design. The internal temperatures can be taken as the design temperature specified for each of the internal spaces.

There is a range of sources for discharge coefficients CIBSE (2005). Manufacturers' data is to be preferred, but in the absence of this, theoretical estimates or experimental measurement may be used. The wind can again introduce uncertainty, due to the effects of external flow (Etheridge 2007a).

Adventitious leakage can be measured on existing buildings. The conventional technique is to use steady pressurisation. This technique requires the use of pressures (typically 50 Pa) that are much higher than those encountered with natural ventilation. The reason for this is to avoid errors arising from wind effects. A new unsteady technique is described in Etheridge (2007a) that allows the leakage to be measured at low pressures.

2.5 Unsteady Flow Models

The main concern of unsteady envelope flow models is the effect of wind turbulence. Slow changes associated with temperature variations can be dealt with by steady flow models. The unsteadiness in the wind manifests itself as relatively rapid fluctuations in the external surface pressures, and these lead to fluctuations in the flow rates through the envelope openings and in the internal pressure. The internal pressure fluctuations are accompanied by changes in density. These changes are small, but they affect a large volume and the associated rates of change of mass can be of similar magnitude to the total flow rates through the openings. Thus, the unsteady flow theory needs to account for two effects, namely, the unsteady behaviour of the openings and the compressibility of the internal air. Unsteady flow models can be distinguished by the assumptions which are made about the flow equation and the continuity (mass conservation) equation. Here, we are only concerned with one type, the *quasi-steady temporal inertia* (QT) model. A more complete summary of models and the theory is given in Etheridge (2000a, b).

2.5.1 Basic Theory for QT Model

Figure 2.6 shows a single cell of volume V (m^3) with one long opening and one sharp-edged opening. The model is formulated by considering mass conservation within the envelope, the flow characteristics of the openings and the pressure differences across the envelope.

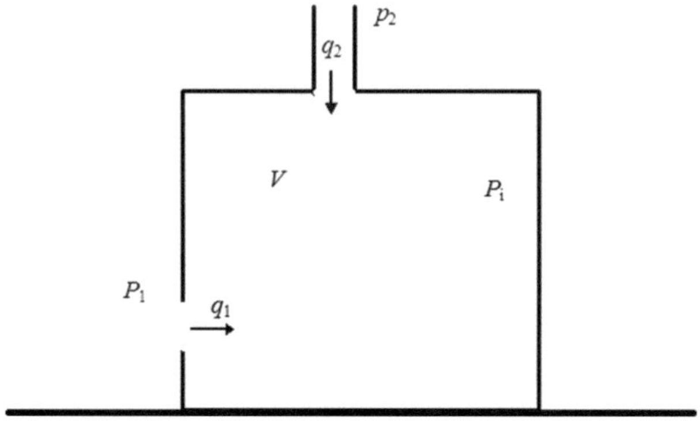

Fig. 2.6 Single cell with two openings (one short, one long)

2.5.1.1 Conservation of Mass

The compressibility effect is relatively straightforward, because the behaviour of the internal air should follow the isentropic relation between pressure and density. It can therefore be accounted for by a modification to the continuity equation. For the conditions which are considered here, the effect of compressibility is usually less important than the effect of inertia, particularly when dealing with instantaneous flows in ducts. However, at high frequencies of fluctuation, compressibility can be important, and since it is relatively easy to model, it should generally be included.

Applying the mass conservation (continuity) equation to the air bounded by the envelope

$$\frac{dm\{t\}}{dt} = \rho_1\{t\}q_1\{t\} + \rho_2\{t\}q_2\{t\} \tag{2.27}$$

Where $m = \rho_i V$ and $\{\}$ denotes a function of time. (2.28)

Strictly speaking, ρ_1 and ρ_2 depend on the direction of flow, but the small differences in density can be ignored on the right-hand side of Eq. (2.27) to give

$$V\frac{1}{\overline{\rho}_i}\frac{d\rho_i\{t\}}{dt} = q_1\{t\} + q_2\{t\}, \tag{2.29}$$

where $\overline{\rho}_i$ denotes the time-averaged internal density.

The behaviour of the bulk of the air contained in the cell can be considered to be isentropic

$$\frac{P_{i0}}{\rho_i^{\gamma}} = C_0, \tag{2.30}$$

where $P_{i0}\{t\}$ denotes the internal pressure at ground level (the variation with height can be neglected in terms of density changes) and C_0 is a constant.

Substituting for ρ_i from Eq. (2.30) into Eq. (2.29) leads to

$$\frac{dP_{i0}\{t\}}{dt} = \frac{\gamma}{V}\overline{P}_{i0}(q_1\{t\} + q_2\{t\}) \tag{2.31}$$

where $\overline{P}_{i0}$ denotes the time-averaged internal pressure.

2.5.1.2 Momentum Equation

The unsteady flow behaviour through the openings is more difficult to treat. In the quasi-steady temporal inertia theory, the pressure difference across an opening is divided into three components associated with momentum change between inlet and outlet, wall friction (with long openings such as stacks) and unsteady inertia. This approach corresponds to application of the principle of linear momentum to the flow through the opening. To account for the first two components, it is assumed that they are given by the steady flow relationships applied at each instant of time – hence, the term "quasi-steady". The third component, the inertia term, is accounted for by evaluating the force (per unit area) required to accelerate an "effective volume" of air which has a velocity equal to the mean velocity through the opening. The effective volume is partly empirical.

Two types of opening are considered. The first is a sharp-edged opening (type 1 or 3), such as an open window or air vent, with a steady discharge coefficient, C_d, equal to about 0.6. The second type is a long duct (type 2), where the discharge coefficient depends on the length/diameter ratio, L/d, and on Reynolds number, Re.

In truly steady flow conditions, the flow characteristic of an opening can be represented by a quadratic relation between the pressure difference across the opening and the flow rate:

$$p_1 - p_i = aq^2 + bq. \tag{2.32}$$

The values of the coefficients a (Pa s^2/m^6) and b (Pa s/m^3) depend on the geometry of the opening and can in principle be determined experimentally in the laboratory or in the field. For certain shapes of opening, they can be determined directly from the geometry (size and shape) of the opening. For a sharp-edged opening (orifice), $b = 0$.

The quasi-steady temporal inertia theory is used to describe the unsteady flow through the opening, with the following flow equation for a sharp-edged opening:

$$p_1\{t\} - p_{i1}\{t\} = a_1 q_1\{t\}^2 + c_1\frac{dq_1\{t\}}{dt}. \tag{2.33}$$

This equation can be obtained from the free streamline theory for steady flow through a sharp-edged orifice by assuming that the streamline pattern is independent of time (Etheridge and Sandberg 1996). The first term on the right-hand side corresponds to the quasi-steady flow assumption (a_1 has the same value as the steady value). The second term is the inertia term. In Eq. (2.33), the pressures are at points away from the opening, where they are not influenced by the flow through it. This means that the inertia term in Eq. (2.33) takes account of the inertia of the air surrounding the inlet and the outlet of the opening. The coefficient c_1 can be expressed in terms of an effective inertia length l_e (m), and the relationship between l_e and the geometry of a circular orifice has been found by comparing the predictions of (34) with experimental data (Etheridge 2000a). This led to the relation

$$l_e = \frac{d}{0.6} \tag{2.34}$$

For the long opening, we have

$$p_2\{t\} - p_{i2}\{t\} = a_2 q_2\{t\}^2 + b_2 q_2\{t\} + c_2 \frac{dq_2\{t\}}{dt}. \tag{2.35}$$

The first two terms on the right-hand side correspond to the quasi-steady flow assumption (the steady values for *a* and *b* are used), and the third term is the inertia term. The inertia coefficient c_2 also depends on the geometry of the opening, but it cannot generally be measured, and some other way of specifying it is needed. The inertia of the inlet and outlet flow is taken to be the same as that for the sharp-edged orifice, to which must be added the inertia of the air within the opening (Etheridge 2000b). For a cylindrical opening of length *L*, the effective inertia length is taken to be

$$l_e = \frac{d}{0.6} + L. \tag{2.36}$$

2.5.1.3 Pressure Difference Across an Opening

The external pressure at height z_1 and time *t* is given by the sum of the hydrostatic pressure and wind pressure (the external density is uniform):

$$P_{E1}\{t\} = P_{E0} - \rho_E g z_1 + p_{w1}\{t\}, \tag{2.37}$$

where P_{E0} denotes the hydrostatic pressure at $z = 0$, ρ_E the external density and p_{wI} the pressure due to the wind. The internal pressure at height z_1 and time *t* is given by (the internal density is uniform and internal air speeds are negligible)

$$p_{I1}\{t\} = P_{I0}\{t\} - \bar{\rho}_I g z_1, \tag{2.38}$$

where P_{I0} denotes the hydrostatic pressure at $z = 0$ and $\bar{\rho}_I$ the time-averaged internal density (the fluctuations of density can be neglected here).

The pressure difference acting across opening 1 is therefore

$$P_{E1}\{t\} - P_{I1}\{t\} = P_{E0} - P_{I0}\{t\} - \Delta\rho.gz_1 + p_{w1}\{t\}, \tag{2.39}$$

where $\Delta\rho$ denotes the density difference between the interior and exterior. Similarly, for opening 2, we have

$$P_{E2}\{t\} - P_{I2}\{t\} = P_{E0} - P_{I0}\{t\} - \Delta\rho.gz_2 + p_{w2}\{t\}. \tag{2.40}$$

2.5.1.4 Equations to Be Solved

For the quasi-steady temporal inertia model, the QT model, there are three simultaneous differential equations to be solved, for the internal pressure difference (P_{E0}–$P_{I0}\{t\}$) and the two flow rates, i.e. mass conservation within envelope:

$$\frac{dP_{I0}\{t\}}{dt} = \frac{\gamma}{V}\overline{P}_{I0}(q_1\{t\} + q_2\{t\}), \tag{2.41}$$

flow through sharp-edged opening

$$c_1\frac{dq_1\{t\}}{dt} = P_{E0} - P_{I0}\{t\} - \Delta\rho.gz_1 + p_{w1}\{t\} - a_1q_1\{t\}^2, \tag{2.42}$$

flow through long opening

$$c_2\frac{dq_2\{t\}}{dt} = P_{E0} - P_{I0}\{t\} - \Delta\rho.gz_2 + p_{w2}\{t\} - aq_2\{t\}^2 - b_2q_2\{t\}. \tag{2.43}$$

The boundary conditions comprise the initial values of the three variables, (P_{E0}–$P_{I0}\{t\}$), $q_1\{t\}$ and $q_2\{t\}$ and the time records of p_{w1} and p_{w2}. The equations can be solved for given boundary conditions using established numerical techniques (e.g. Matlab was employed in Etheridge (2000a, b)). It is useful to express the equations in nondimensional form prior to solving them, because the terms retain roughly the same size irrespective of whether they apply to a full-scale building or a small wind tunnel model.

2.5.2 *Unsteady Flow Models in Design*

The direct use of the QT model in design is limited by the fact that it is necessary to know the *simultaneous* variation of the pressure at each opening. Such data is time-consuming to obtain, even from wind tunnel models, and as a consequence, it is generally not available to the designer. Nevertheless, it is desirable to know whether unsteady effects are important and, if they are, to take some account of them. The studies described in Etheridge (2000a, b) showed that this can best be done by making use of results of theoretical calculations expressed in nondimensional form.

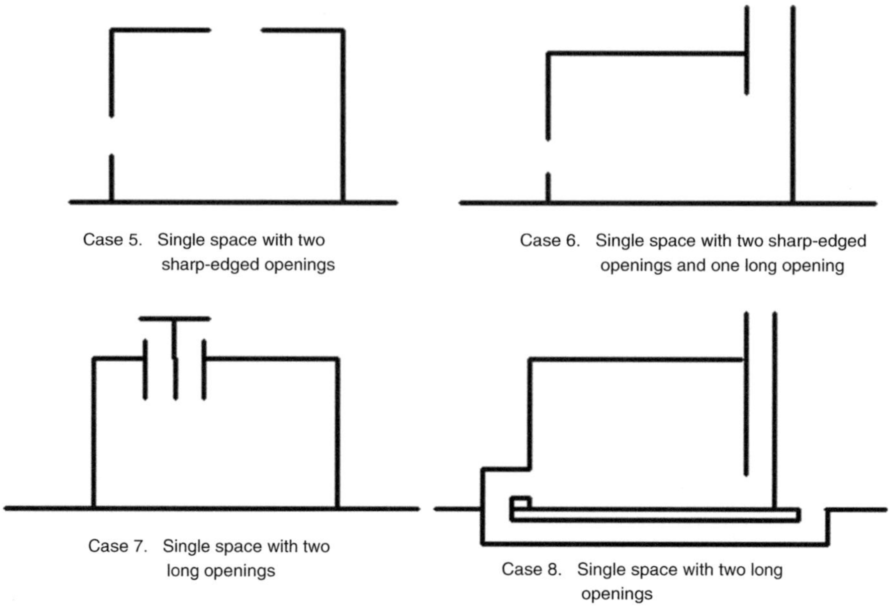

Fig. 2.7 Cases of buildings or spaces with two dominant openings

In this way, some general results can be applied to the design process. This is particularly true for mean (i.e. time-averaged) flow rates, which are of primary interest to the designer. The other effect of unsteadiness relates to flow reversal in stacks, and this too can be treated in the same way, albeit with less generality.

In the following, some results of the studies are described. The studies were limited to the case of single-cell spaces with two dominant openings with constant discharge coefficients. The limitation to two dominant openings is not as restrictive as it may sound, since there are many cases which can be included in this category, as shown in Fig. 2.7.

2.5.3 Important Nondimensional Parameters

From the analysis of the results obtained in the study, the nondimensional parameter $W/\sigma_{\Delta Cp}$ was found to be of major importance. This parameter is a measure of the relative sizes of the steady pressure difference (due to wind *and* buoyancy) and the fluctuating component of the pressure difference (due to wind). W (−) is defined by

$$W = Ar\left(\frac{z_2}{H} - \frac{z_1}{H}\right) + \left(\frac{\overline{C_{p1}} - \overline{C_{p2}}}{2}\right), \tag{2.44}$$

where the first term is due to buoyancy and the second term is the difference between the mean wind pressure coefficients. $\sigma_{\Delta \mathrm{Cp}}$ (−) is the standard deviation of the instantaneous difference between the surface wind pressure coefficients. The Archimedes number Ar (−) is defined by

$$Ar = \frac{\Delta \rho g H}{\rho U^2}. \tag{2.45}$$

Clearly, as $W/\sigma_{\Delta \mathrm{Cp}}$ becomes large, unsteadiness will have a smaller effect and vice versa.

Two other nondimensional parameters which indicate the importance of compressibility and inertia are denoted by D (−) and F (−), respectively, and are defined by

$$D \equiv \frac{2C_{d1}A_1 H}{V} \frac{nc^2}{\gamma U^2}, \tag{2.46}$$

$$F \equiv \frac{H}{l_e}. \tag{2.47}$$

Both parameters are inverse ones, i.e. the lower the value, the larger the effect. Low values of D and F do not necessarily mean that compressibility and inertia effects will be significant. It is also necessary for the nondimensional frequencies of the pressure fluctuations to be sufficiently large.

Increasing the volume, V, of a building will decrease D and increase compressibility effects, if all other properties are unchanged. In physical terms, the larger the building volume, the greater is the change in the mass of that volume arising from a change in internal pressure. This has an influence on the flow rates through the openings, via the continuity equation. The appearance of c (m/s) (speed of sound) in the parameter D stems from the inclusion of compressibility. It does not mean that account is taken of the time taken for pressure waves to travel through the volume. The theory assumes that such times are negligible; i.e. internal pressure changes occur instantaneously throughout the volume. With very large buildings, this assumption becomes increasingly invalid.

The parameter F is simply a measure of the relative size of the effective length, l_e, of an opening. The greater the effective length, the lower the value of F and the greater is the inertia (mass) of the air flowing through the opening. The greater the inertia, the greater is the resistance to change in flow rate and direction.

2.5.4 Unsteady Effects on Mean Flow Rates and a Method for Estimating the Effects

The main effects of unsteadiness on mean flow rates for Case 5 in Fig. 2.7 are summarised by the solid line in Fig. 2.8. The line is a fit to calculated results, but the variation about the line is not great (typically less than 10 %). The nondimensional total fresh air flow rate (the flow rate divided by C_dAU) is plotted against the pressure parameter $W/\sigma_{\Delta Cp}$. The dashed line shows the results of the steady envelope model.

For $W/\sigma_{\Delta Cp} > 1.0$, the effect of unsteadiness is small, and in the context of design, the steady calculation method should suffice for $W/\sigma_{\Delta Cp} > 0.4$. When $W/\sigma_{\Delta Cp} < 0.4$, it may be preferable to employ the unsteady calculation method, but the requirement for simultaneous records of pressure fluctuations is very demanding. If such records are not available, the nondimensional graph of Fig. 2.8 can be used to give reasonable estimates much more easily. The main problem is to specify $\sigma_{\Delta Cp}$. If the standard deviations of the pressure coefficients are known, σ_1 and σ_2, they can be used to estimate $\sigma_{\Delta Cp}$. Tabulated data for σ_1 and σ_2 is available, although not to the same extent as mean pressure coefficients. It is worth noting that high temperature differences (large Ar), high ($C_{p1}-C_{p2}$) and/or low wind speeds lead to values

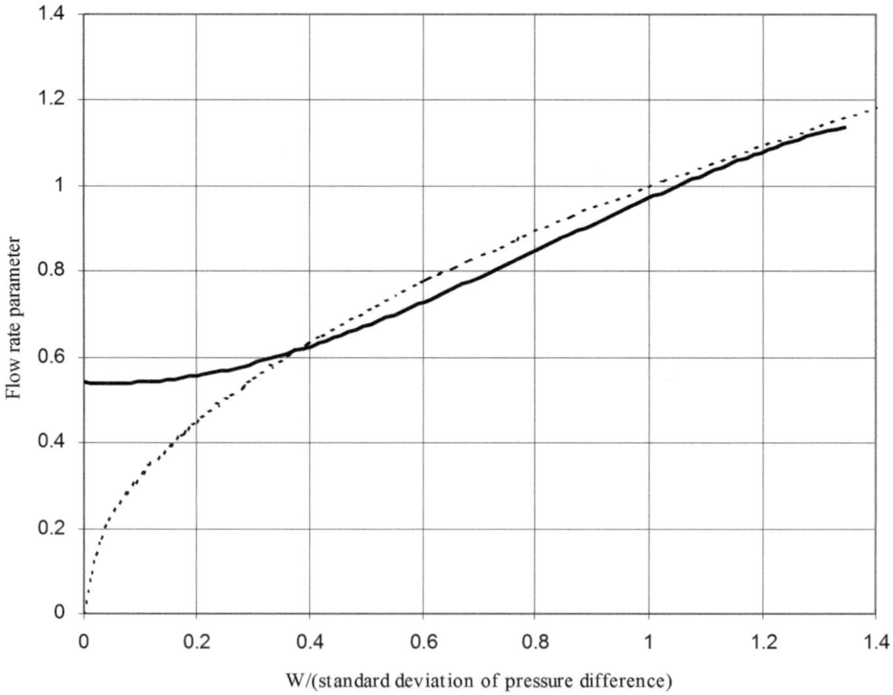

Fig. 2.8 Comparison of nondimensional ventilation rates for steady (*dashed line*) and QT (*solid line*) models (Adapted from (Etheridge 2000a), Elsevier)

of $W/\sigma_{\Delta Cp}$ which are much greater than 0.4. For many situations, therefore, an unsteady calculation is not needed for *mean* flow rates.

2.5.5 Instantaneous Flow Rates

Not surprisingly, the parameter $W/\sigma_{\Delta Cp}$ is also of considerable importance to the instantaneous flow rates in stacks. However, because we are dealing with instantaneous flow rates, it is not possible to summarise the results in such a general way as for mean flows, partly because the detailed nature of the instantaneous pressures is important.

As far as the designer is concerned, the main feature of interest is likely to be in departures of the flows from design specification. Fluctuations in the *magnitudes* of flow rates must be expected in natural ventilation, and it is the *direction* of flows which are of more importance. Flow reversal in stacks and ducts is undesirable, and it may be necessary to estimate the likelihood of this occurring. An important feature is to what extent the use of ducts with their increased inertia reduces the occurrence of flow reversal. This has been investigated using simple hypothetical gusts and recorded time histories.

Figure 2.9 shows the volume of flow reversal for a given gust strength at typical winter design conditions. Three curves are shown corresponding to the three

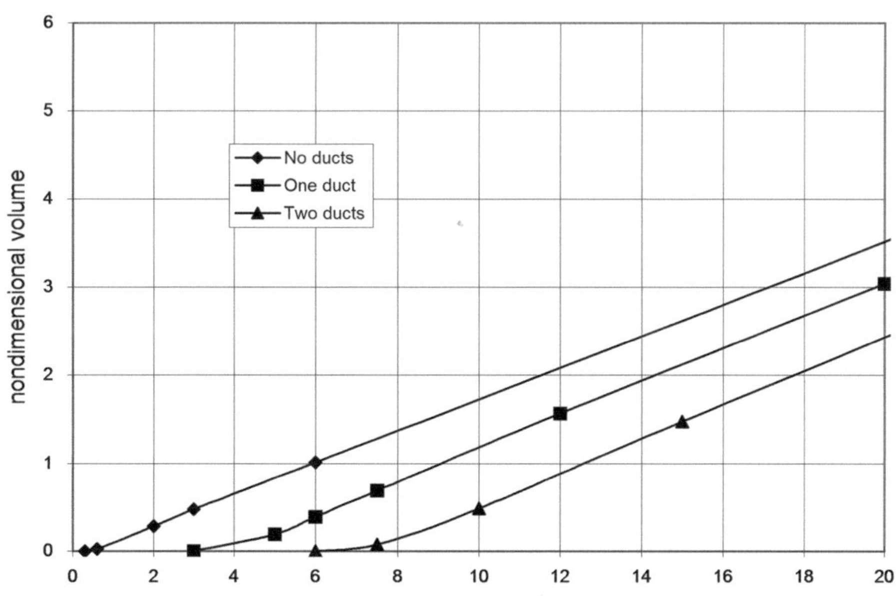

Fig. 2.9 Plot of nondimensional reversal volume against nondimensional gust strength for Case 5 (*no ducts*), Case 6 (*one duct*) and Case 8 (*two ducts*) (Adapted from (Etheridge 2000b), Elsevier

different opening configurations, namely, Cases 5, 6 and 8 in Fig. 2.7. The nondimensional reversal volume is defined by

$$\frac{\text{volume passed during period of flow reversal}}{C_d A H} \tag{2.48}$$

and the nondimensional gust strength by

$$\frac{C_{pg}\tau_g U}{H} \tag{2.49}$$

where C_{pg} (−) and τ_g (s) are the amplitude and duration of the gust. For these calculations, the gust was taken as a sinusoidal variation of the external pressure coefficient at one of the openings. The main point of interest is the effect of the ducts, and it can be seen that they lead to a clear increase in the resistance to flow reversal. For example, the gust strength required to cause flow reversal increases from about 0.8 to 3.0 with one duct and to 6.0 with two ducts.

In reality, the effect of ducts is rather more complex and depends on the detail of the pressure fluctuations. Figure 2.10 shows results obtained for real pressure records in the form of a plot of the number of times the reversal volume exceeded

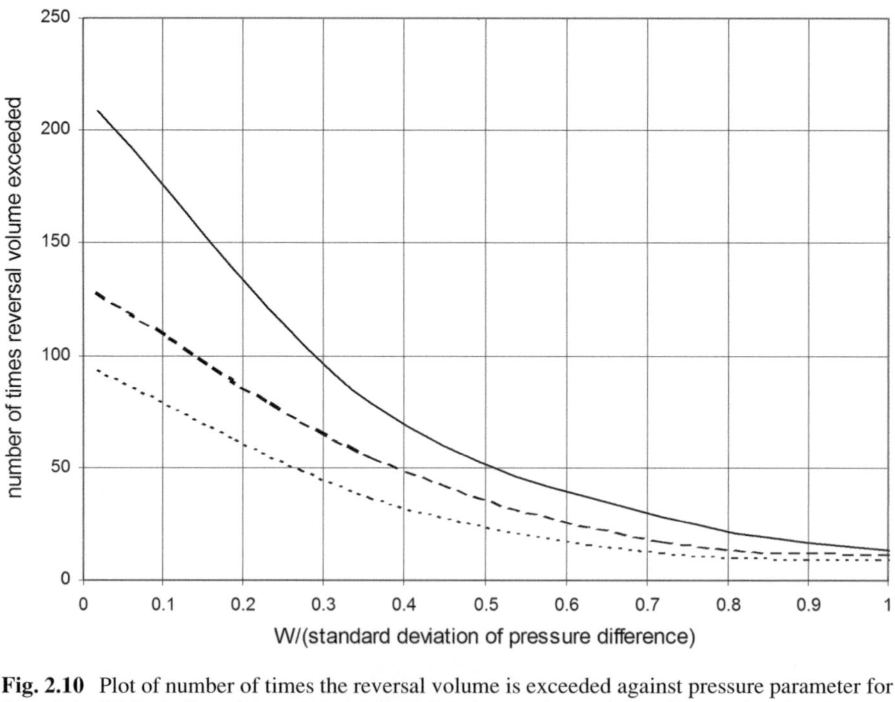

Fig. 2.10 Plot of number of times the reversal volume is exceeded against pressure parameter for Case 5 (*solid line*), Case 6 (- - - -) and Case 8 (··········), using real time-history data for pressure fluctuations (Adapted from (Etheridge 2000b), Elsevier)

the volume of the duct against $W/\sigma_{\Delta Cp}$, again using the three opening configurations. Flow reversal occurs for all configurations, and this is a reflection of the chosen pressure record, which has long periods of adverse pressure differences. The avoidance of flow reversal can only be guaranteed by siting the openings such that $W/\sigma_{\Delta Cp}$ exceeds a value of about 1.5 for all wind directions. This emphasises the need to achieve sites with a large value for $(C_{p1}-C_{p2})$. However, even in the chosen case, the addition of ducts gives clear benefits.

2.6 Conclusions

Steady envelope flow models provide a convenient and simple tool for investigating envelope flow rates at both the initial stage of design and at the later stage. In the early stage, they provide a means for determining the size and position of openings to achieve a specified flow pattern (magnitude and direction) through the external envelope. However, the use of envelope flow models requires knowledge of the surface wind pressure distribution, the discharge coefficients of the openings and the internal density distribution. Conventional models are essentially limited to small openings, where the still-air discharge coefficient can be applied to individual openings and where the opening does not affect the external pressure distribution.

Unsteady envelope flow models require substantially more data, in the form of simultaneous records of external wind pressure fluctuations. For this reason alone, they are unlikely to find common use in design. However, the models do allow investigations of unsteady stack flows, e.g. their propensity to flow reversal. The use of the models may therefore be justifiable when stacks or chimneys form a crucial part of the design. For more general usage, the challenge is to formulate the results of calculations in a nondimensional form that has a wide range of application.

Acknowledgement The permission of CIBSE to reproduce Fig. 2.5 and to quote results from CIBSE (2005) is gratefully acknowledged.

References

Chiu Y-H, Etheridge DW (2002) Calculations and notes on the quadratic and power law equations for modelling infiltration. Int J Vent 1(1):65–77

CIBSE (2005) Natural ventilation in non-domestic buildings, Applications Manual AM10:2005. The Chartered Institution of Building Services Engineers, London

Etheridge DW (2000a) Unsteady flow effects due to fluctuating wind pressures in natural ventilation design – mean flow rates. Build Environ 35(2):111–133

Etheridge DW (2000b) Unsteady flow effects due to fluctuating wind pressures in natural ventilation design – instantaneous flow rates. Build Environ 35(4):321–337

Etheridge DW (2002) Nondimensional methods for natural ventilation design. Build Environ 37:1057–1072

Etheridge DW (2007a) External wind effects on flow through small openings and leakage measurement. COE International Advanced School on Environmental Wind Engineering, COE-IAS4, Soongsil University, Seoul

Etheridge DW (2007b) Scale modelling of natural ventilation. COE International Advanced School on Environmental Wind Engineering, COE-IAS4, Soongsil University, Seoul

Etheridge DW (2007c) Design procedures for natural ventilation. COE International Advanced School on Environmental Wind Engineering, COE-IAS4, Soongsil University, Seoul

Etheridge DW, Sandberg M (1996) Building ventilation: theory and measurement. Wiley, Chichester

Li Y, Delsante A, Chen Z, Sandberg M (2000) Some examples of solution multiplicity in natural ventilation. In: Proceedings of RoomVent. Elsevier, Kidlington

Linden PF, Lane-Serff GF, Smeed DA (1990) Emptying filling boxes – the fluid mechanics of natural ventilation. J Fluid Mech 212:309–335

Ohba M (2007) Study on predicting wind-driven cross ventilation flow rates and discharge coefficients based on Local Dynamic Similarity Model. COE International Advanced School on Environmental Wind Engineering, COE-IAS4, Soongsil University, Seoul

Orme M (1999) Applicable models for air infiltration and ventilation calculations, AIVC TN 51. Air Infiltration and Ventilation Centre, Coventry

Chapter 3
Ventilation Flow Structure and High-Precision Ventilation Network Model

Masaaki Ohba

Abstract Understanding of ventilation flow structure is important to utilize the cross-ventilation for improving indoor environment in hot and humid rooms. The pressure loss at the opening depends on the contraction of ventilation air flow and on static pressure loss consumes for the production of turbulent kinetic energy. The local dynamic similarity model (LDSM) is capable to predict the ventilation flow rate with better accuracy. LDSM has superiority over the conventional orifice flow model when discharge coefficients decreased due to changes of wind directions. A simulation study by the ventilation network model is described for performing on cooling loads in a detached house using by active operation of windows.

Keywords Cross-ventilation • Wind tunnel experiment • Large eddy simulation • Stream tube • Local dynamic similarity model • Ventilation performance • Ventilation flow rate • Gross coverage ratio • Cooling load • Ventilation network model

3.1 Introduction

Natural ventilation is an energy-efficient technology that is adopted to reduce energy consumption for heating and cooling of buildings (Per Heiselberg 2002; Nonaka et al. 2009). The cross-ventilation is defined as the large wind-enforced ventilation, which is composed of the natural ventilation with stack ventilation. In order to effectively promote the utilization of cross-ventilation, it is important to understand the structures of the ventilation flow near the openings. Ventilation is a phenomenon of complicated turbulent flow, which exhibits reversible and irreversible changes of energy between dynamic pressure and static pressure associated with extreme deformation of air flow near openings. It is difficult to experimentally reproduce the spatial distribution of these pressures, and this hinders the elucidation of the problem.

M. Ohba (✉)
Tokyo Polytechnic University, 1583 Iiyama, Atsugi, Kanagawa 243-0297, Japan
e-mail: ohba@arch.t-kougei.ac.jp

Y. Tamura, R. Yoshie (eds.), *Advanced Environmental Wind Engineering*,
DOI 10.1007/978-4-431-55912-2_3

There have been many studies regarding mechanics of cross-ventilation (Kurabuchi et al. 2004). Shohda and Sekine observed that the pressure loss coefficient of opening exposed to external wind is different from that in stagnant surrounding condition and varies according to the wind direction angle (Shohda and Sekine 1961). Ishihara introduced influential factor on cross-ventilation that is defined as the observed pressure loss coefficient of multiple openings connected in series divided by the sum of pressure loss coefficient in each opening (Ishihara 1969). He suggested that influential factor tends to take a value smaller that one in most cross-ventilation situation. Vickery and Karakatsanis studied applicable range of the orifice flow theory to the cross-ventilation by comparison of measured and calculated flow rate in wind tunnel experiment (Vickery and Karakatsanis 1987). They found the orifice flow theory with fixed discharge coefficient and wind pressure distribution on a sealed building model could evaluate flow rate within 10 % error when incident flow angle to the opening is smaller than 45° and the porosity is smaller than 23 %. They found that the theory overestimated the flow rate outside this condition as a result of the change of discharge coefficient. Aynsley proposed to apply resistance-based flow theory rather than orifice flow theory to account for the effect of airway resistance, dynamic flow resistance, and friction flow resistance (Aynsley 1988).

Murakami, Kato et al. stressed the importance of the remaining dynamic pressure to promote ventilation rate that reaches the leeward opening before being dissipated at the downstream side of the windward opening (Murakami et al. 1996). They also suggested that cross-ventilation phenomena should be regarded as diverging and converging transport of power such as multiple pipe flow networks. Sandberg introduced a geometric parameter named catchment area to illustrate dependence of porosity to discharge coefficient (Sandberg 2002).

Although these works are useful for understanding some aspects of the cross-ventilation phenomena, none of them has been established so far to replace the conventional orifice flow theory in the practical application scene. Sawachi et al. have confirmed large change of discharge coefficient affected by wind direction angles using actual-size building model situated in a large wind tunnel and stressed the importance of its accurate evaluation for accurate prediction of ventilation flow rate (Sawachi 2002).

This chapter describes the ventilation flow structure through the use of wind tunnel experiments and CFD. A new ventilation model of local dynamic similarity model (LDSM) which has been proposed by authors (Kurabuchi et al. 2002b, 2005a) based on the CFD analysis is also introduced and is capable to predict the ventilation flow rate with better accuracy than that by the conventional orifice model. Finally, a simulation study by the ventilation network model is described for performing on cooling loads in a detached house using active operation of windows (Tsukamoto et al. 2009, 2013).

3.2 Comparison of LES Calculation and Experimental Results Inside and Outside a Cross-Ventilation Model

A cross-ventilation model in the wind tunnel experiments was used to measure the outdoor and indoor airflow and the flow stream in the vicinity of the openings. There were two openings on the two sides of the walls to allow the flows coming in and out for cross-ventilation. The split-film probes were used to measure the directional velocities and turbulence levels inside and outside the cross-ventilation model.

Figure 3.1 compares the wind velocity vectors in the vertical symmetrical cross-section with wind direction set to 0°. Under this condition, it is known that ventilation air flow after flowing into the room turns steeply downward due to the transport of downward momentum immediately before the opening and the pressure gradient caused by the circulating flow developed at the lower portion of the building. The result of the large eddy simulation (LES) reproduces well the flow features (Kurabuchi et al. 2000, 2002a).

Figure 3.2 compares the distribution of turbulent kinetic energy in the same cross-section. Turbulent kinetic energy increases at the front edge of the roof surface where the flow is separated and in the ventilation air flow area after it flows into the room, and the values correspond well with each other. They are relatively low in front of the building where the approach flow impinges.

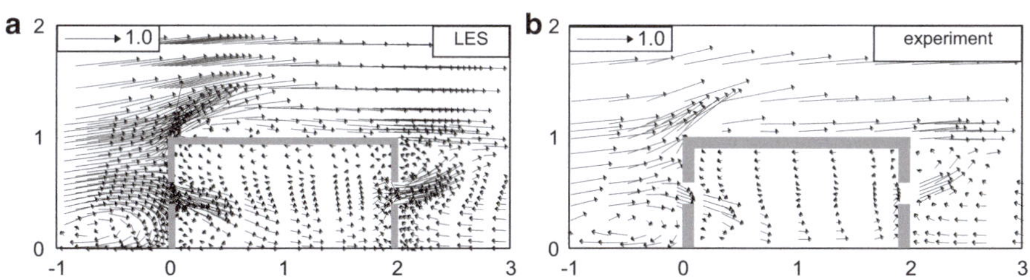

Fig. 3.1 Comparison of velocity vector distributions inside and outside a cross-ventilation model: (**a**) LES, (**b**) experimental measurements

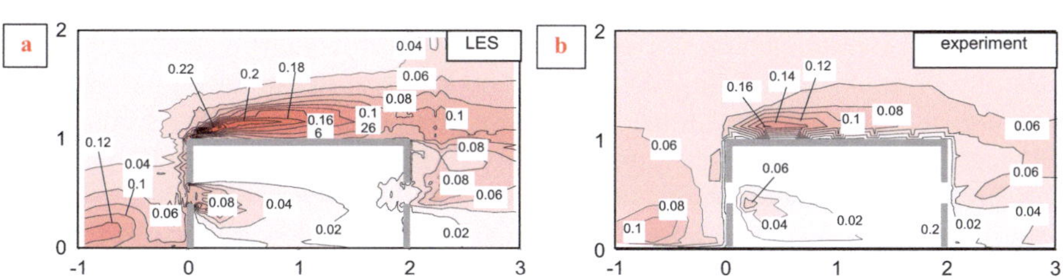

Fig. 3.2 Comparison of turbulent kinetic energy inside and outside a cross-ventilation model: (**a**) LES, (**b**) experimental measurements

Fig. 3.3 Comparison of mean wind pressure coefficients: LES and experimental measurements

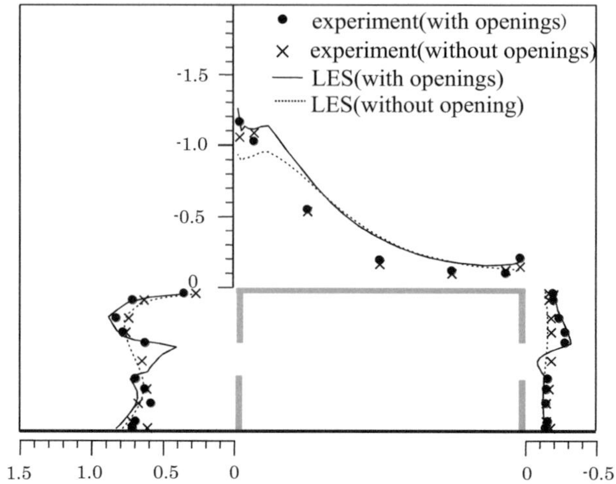

Figure 3.3 represents the comparison of calculation results about wind pressure coefficients using LES and experiment including the case where openings are not present. The agreement between experimental data and LES results was very good. In particular, in case there was the opening, wind pressure coefficient was slightly increased on the upper portion of the opening compared with the case of no opening, and the results of the experiment were reproduced very well.

One of the features of the airflow during cross-ventilation is extreme downfall of inflow. To elucidate the cause, the momentum balance along the central axis was valuated, which passes through the center of the opening. The equation of time-averaged velocity component in vertical direction at the plane of symmetry can be approximately expressed as in the Eq. (3.1):

$$\frac{\partial U_3}{\partial t} \cong -\frac{\partial P}{\partial x_3} - \frac{\partial U_1 U_3}{\partial x_1} - \frac{\partial \overline{u_1 u_3}}{\partial x_1} - \frac{\partial U_3 U_3}{\partial x_3} - \frac{\partial \overline{u_3 u_3}}{\partial x_3} \tag{3.1}$$

where U_i and u_i represent time-averaged velocity component and fluctuation velocity component in x_i direction, respectively, and the superscribed bar indicates the time-averaged value.

Figure 3.4 shows the average airflow field around the central axis and profile of vertical velocity component U_3 (positive in upward direction). Longitudinal profiles of each term in the Eq. (3.1) are given in Fig. 3.5. In the far region from the opening, vertical velocity is turned from positive to negative, and the second term $\partial U_1 U_3/\partial x_1$ is turned to positive. When the airflow approaches the opening, acceleration occurs and the second term has a high positive value. The cause to deflect the flow downward may be the negative contribution in Eq. (3.1), which offsets the above action. According to Fig. 3.5, the pressure gradient $\partial P/\partial x_3$ opposes from the point -1.5 to around the opening except in the vicinity of opening

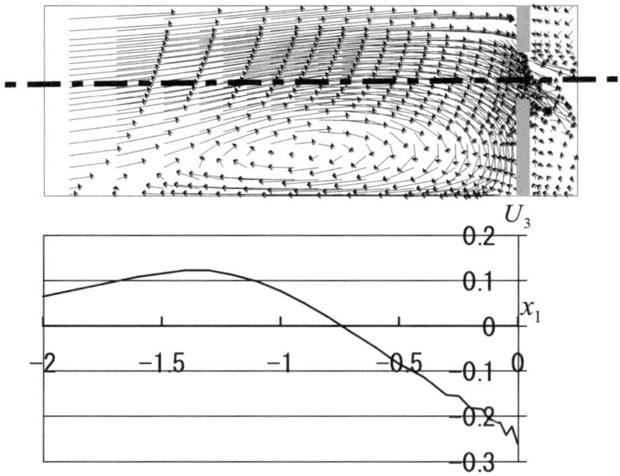

Fig. 3.4 Flow field near windward opening and longitudinal profile of U_3 at central axis of opening

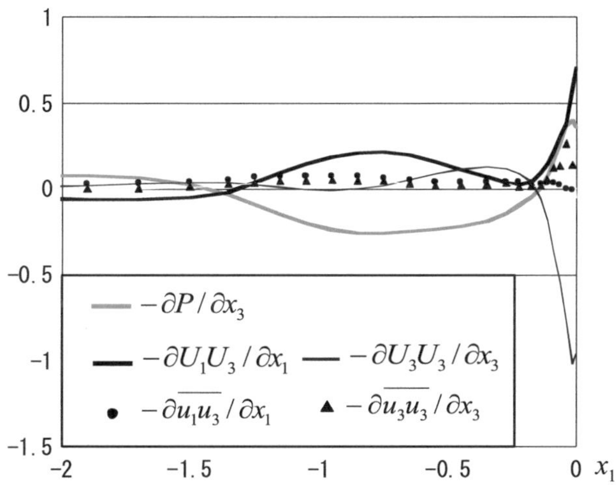

Fig. 3.5 Longitudinal balance of U_3 transport equation at central axis of opening

where only momentum transport by vertical velocity component takes negative value. It may conclude the primary cause of the downfall lies in the pressure gradient due to the recirculation at the lower portion on windward surface.

3.3 Conservation of Total Pressure and Structure of Ventilating Flow

Based on the fact that LES accurately reproduces the experimental result, the structure of ventilation air flow from the analysis of calculation results was elucidated (Kurabuchi et al. 2005b). First, passive markers are set out from the opening's

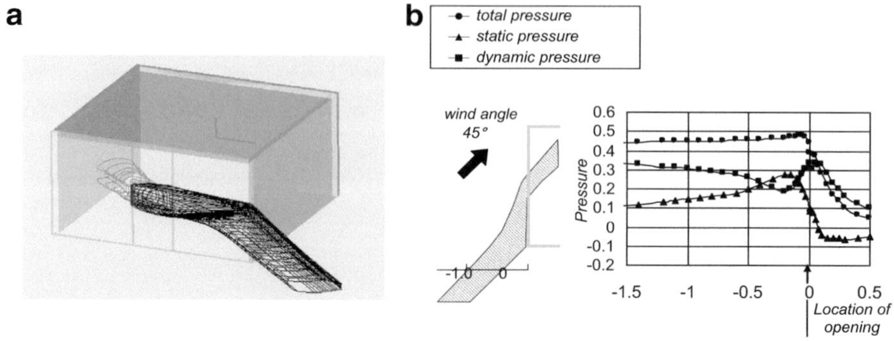

Fig. 3.6 Shape of stream tube and the streamwise change of pressure in the vicinity of opening for wind angle of 45°: (**a**) shape of stream tube, (**b**) streamwise change of pressure

upstream section. By tracing these trajectories upstream and downstream, stream tube shapes before and after the passing at the opening were determined. When the wind direction is other than 0°, the stream tube contacts the wall surface before it reaches the opening, as shown by the result of the case in Fig. 3.6a, where the wind direction is set to 45°. It is turned to a flow along the wall surface and reaches the opening. This means that in most cases the ventilation air flow may be approximated by a wall jet or boundary layer flow before it flows into the opening.

The flow rate weighted average values of total pressure, static pressure, and dynamic pressure in each cross-section of the stream tube were calculated. The changes of these values together with the shape of the stream tube are shown in Fig. 3.6b. From this figure, it is apparent that static pressure and dynamic pressure show extreme changes before reaching the opening when the wind direction is 45°, while the total pressure, i.e., the sum of the two, is almost constant, and pressure loss is low in the process where the wind flows along the windward wall surface.

Figure 3.7 shows the spatial distribution of production of turbulent kinetic energy Pk as defined by Eq. (3.2) when the wind direction is set to 45° and 60°, respectively. When the wind direction is 45°, there is no production of turbulent kinetic energy at the windward corner:

$$P_k = -\overline{u_i u_j}\frac{\partial U_j}{\partial x_i} \tag{3.2}$$

When the wind direction is 60°, significant production is recognized, and this suggests that turbulent kinetic energy is produced in the process where the air flow passes through this portion and total pressure loss occurs. In the shape of the stream tube, the flow is separated at the windward corner and the flow reattaches again to the wall, and total pressure is decreased in this process.

Figure 3.8 shows the changes of total pressure, wind pressure, and room pressure at the opening where the approaching flow angle is changed. Until the wind direction reaches 45°, total pressure at the opening is constant. Total pressure of

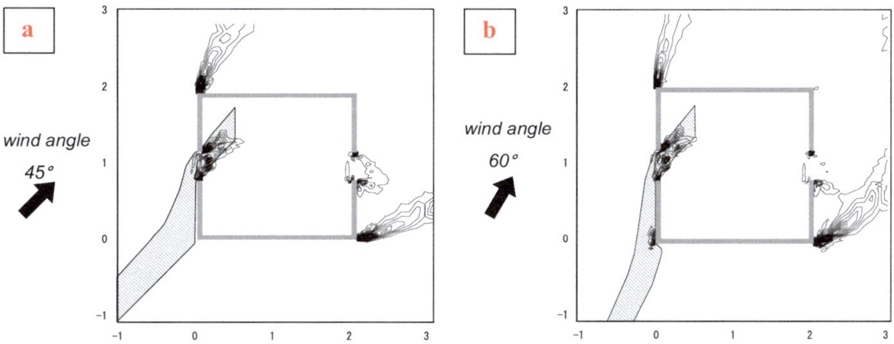

Fig. 3.7 Shape of stream tube and distribution of production of turbulent kinetic energy: (**a**) wind angle of 45°, (**b**) wind angle of 60°

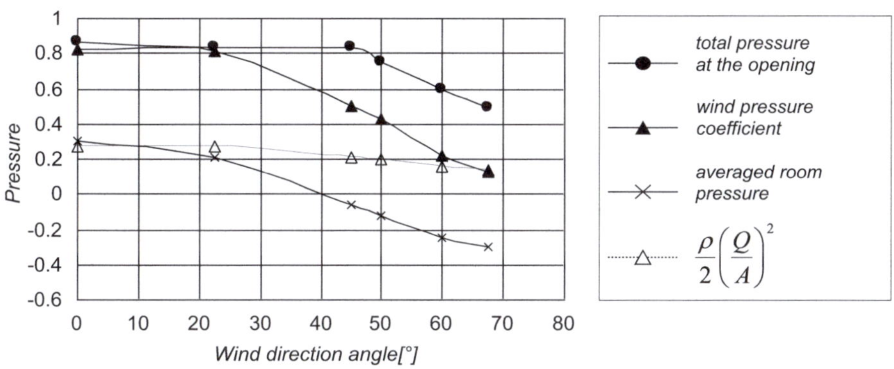

Fig. 3.8 Evaluated pressure at different wind direction angles

an approach flow at approximately the same height as the opening is maintained. When the wind direction exceeds 45°, air flow is separated at the windward corner, and the total pressure is greatly decreased. However, the wind pressure is gently decreased as being apart from the wind direction 0°. When the air flow is not separated on the upstream side of the stream tube, the total pressure at the opening is kept almost constant regardless of the wind direction.

3.4 Concept of Local Dynamic Similarity Model

Kurabuchi et al. (2002b, 2004, 2005b) have proposed a new ventilation model for evaluating the discharge coefficient and flow angle at an inflow opening for cross-ventilation based on the analysis of flow structure from the large eddy simulation. This model is based on the idea that the cross-ventilation flow structure in the

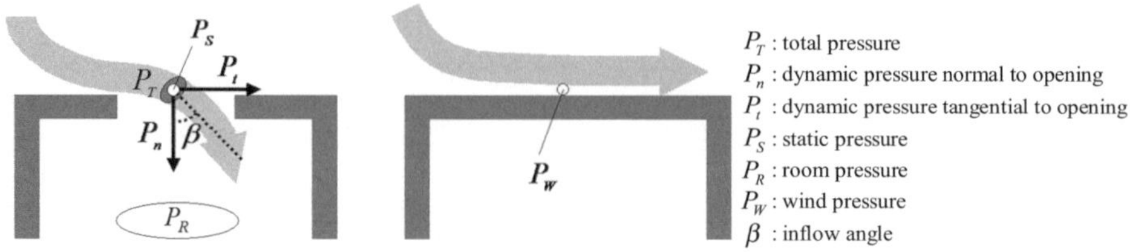

Fig. 3.9 Definition of pressures in vicinity of inflow opening

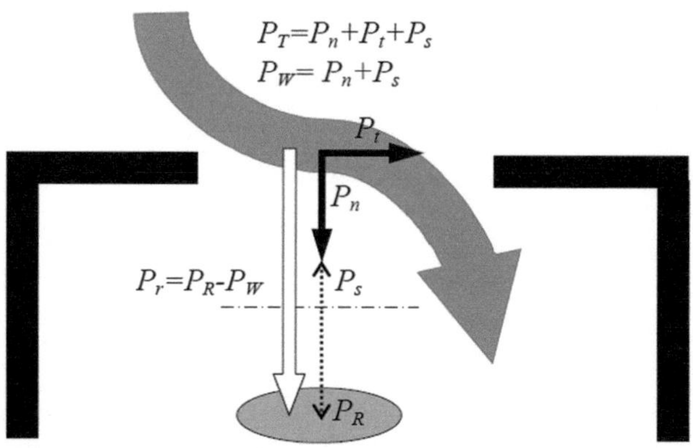

Fig. 3.10 Dynamic similarity in vicinity of inflow opening

vicinity of an inflow opening creates dynamic similarity under the condition that the ratio of cross-ventilation driving pressure to dynamic pressure of cross-flow at the opening is consistent.

Figure 3.9 defines the pressures in the vicinity of the inflow opening, and Fig. 3.10 shows the dynamic similarity in that of the inflow opening. The discharge coefficient (Cd), the inflow angle (β), and the dimensionless room pressure (PR*) are defined as follows:

$$C_d = \sqrt{\frac{P_n}{|P_R - P_W|}} \tag{3.3}$$

$$\beta = \tan^{-1}\sqrt{\frac{P_t}{P_n}} \tag{3.4}$$

$$P_R{*} = \frac{P_R - P_W}{P_t} \tag{3.5}$$

The dimensionless room pressure (P_R^*) represents the ratio of cross-ventilation driving pressure $|P_R - P_W|$, and interfering cross-flow dynamic pressure P_t. P_R* for an inflow opening is always negative. The proposed model supposes that the discharge coefficient (C_d) and the inflow angle (β) are uniquely determined when P_R* is determined.

Then, the ventilation flow rate can be calculated by Eq. (3.6):

$$Q = C_d A \sqrt{\frac{2}{\rho} |P_r|} \tag{3.6}$$

Total pressure P_T at an inflow opening is $P_n + P_t + P_S$, as shown in Fig. 3.10. It is assumed that the $P_n + P_S$ is approximately equal to wind pressure P_W, even when ventilation flow is occurring. Thus, P_t can be evaluated as follows:

$$P_t = P_T - P_W \tag{3.7}$$

Establishment of the local dynamic similarity model may require the following conditions:

- The opening shape has geometrical similarity.
- There is no distribution of Pt and PW inside the opening.
- The direction of tangential flow of the approach flow with respect to the opening is constant.
- The opening is positioned on a wall surface that is sufficiently large with respect to the opening.
- There is no wall to hinder the diffusion of incoming air flow near the opening on the room side.

In the case when the air flow acts around the opening, this can be represented only by the three pressure values extracted above. It is expected that similarity of the flow can be established without depending on the shape of the building, position of the opening, approaching flow angle, etc. This concept may be called as local dynamic similarity in the sense that dynamic similarity exists not in total flow field but only in the vicinity of the opening. In this case, a pair of two dimensionless values prepared from three pressure values represents a specific air flow condition. If these values correspond to each other at one by one, it may be deduced that when one of them is determined, the other is automatically determined.

3.5 Evaluation of Ventilation Performance for Various Types of Inflow Openings

3.5.1 Ventilation Performance Expression for Openings

An approximate expression was derived from the results of ventilation performance evaluation for various types of inflow openings. Figure 3.11 shows the approximate

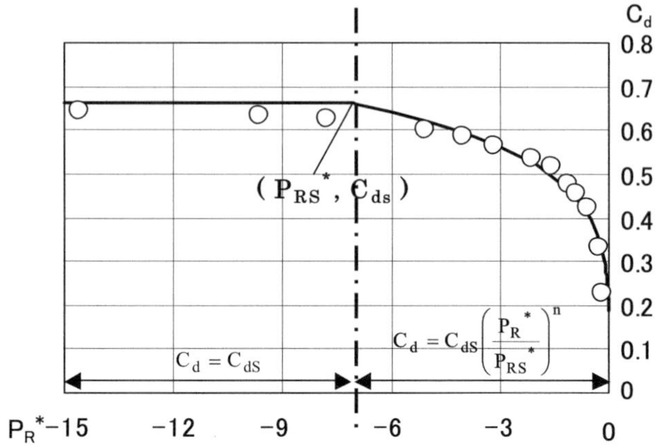

Fig. 3.11 Example of ventilation performance expression for basic opening

expression for the basic opening. These formulae can express C_d as a function of P_R*:

$$C_d = C_{ds}\left(\frac{P_R*}{P_{RS}*}\right)^n \qquad (P_{RS}* \le P_R*) \tag{3.8}$$

$$C_d = C_{d\ S} \qquad (P_R* \le P_{RS}*) \tag{3.9}$$

In the equations, C_{ds} is a basic discharge coefficient. $P_{RS}{}^*$ is P_R* when C_d is equal to C_{ds}. The power exponent n is determined according to the shape of the curve $P_R* - C_d$. The ventilation performance for various types of inflow openings can be easily evaluated by the three parameters C_{ds}, P_R*, and n.

3.5.2 Ventilation Performance Database for Various Types of Inflow Openings

Table 3.1 indicates the details of the openings studied for four different opening shapes: simple opening, louver window, single pivoted window, and single outrigger window. The ventilation performance was done by using a new measurement system ("system for evaluating ventilation performance of openings") (Endo et al. 2009). Using the ventilation performance evaluation method, the relationships between P_R* and C_d for various opening shapes were systematically investigated.

Table 3.2 summarizes the measurement results with previous research data of C_{ds}. The values of C_{ds} for various types of inflow openings are in good agreement with the previously obtained C_{ds} (AIJ 1970). A ventilation performance database for various types of inflow openings is also shown in Table 3.2. By utilizing the database, it is possible to accurately predict the ventilation flow rate.

Table 3.1 Schematic drawing of various types of inflow openings

Opening	shape	Dimension
simple openings	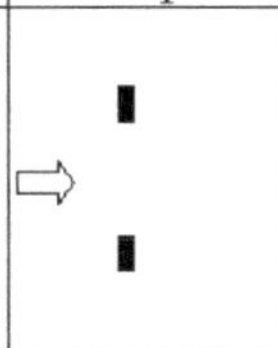	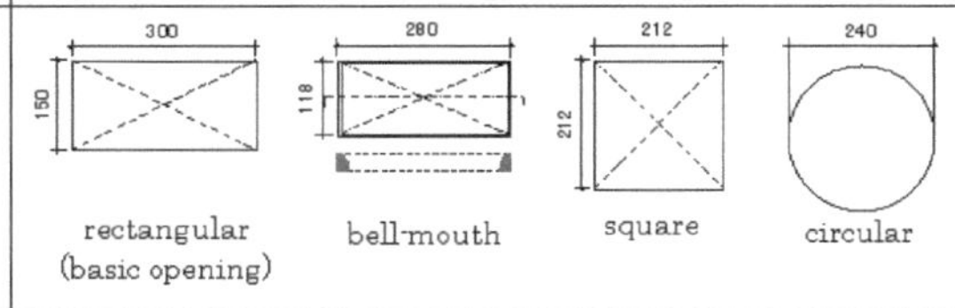
louver windows	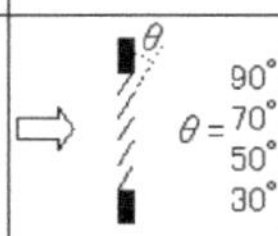	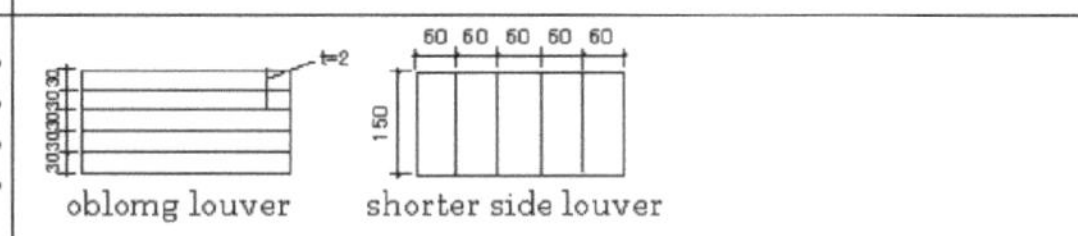
single pivoted windows	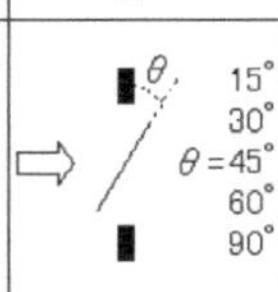	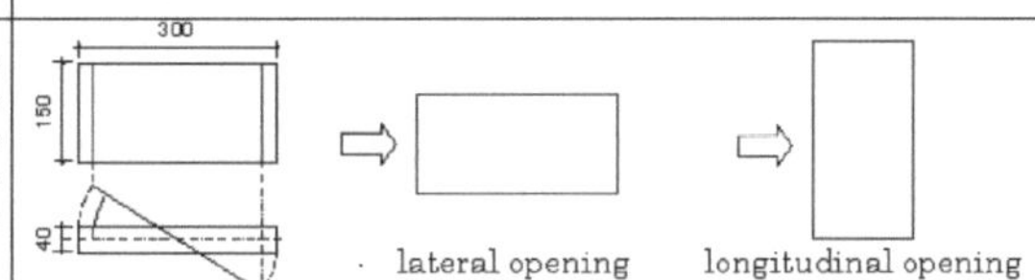
single outrigger windows	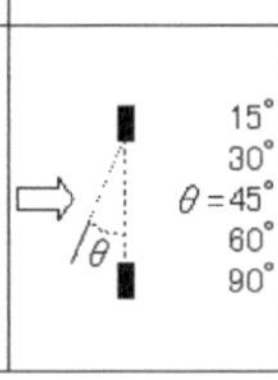	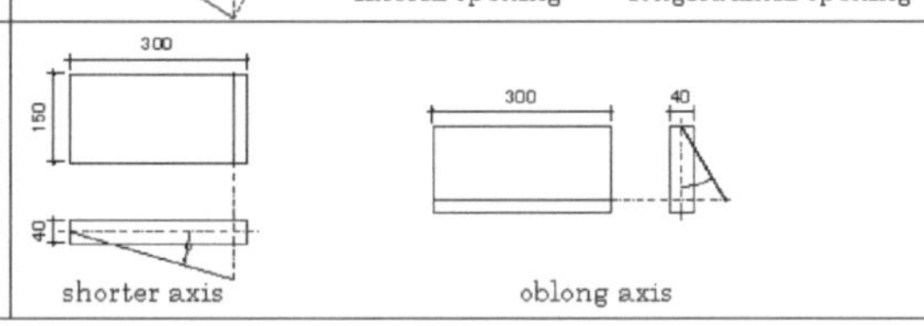

3.5.3 Ventilation Performance for Full-Sized Window Frame

The ventilation characteristics for a full-sized small opening with louvers were investigated by the system for evaluating ventilation performance of openings. Figure 3.12 shows the specifications of the window and the discharge coefficient curves. The detailed features of the window were very complicated. There was no substantial difference between the full-sized window and the small-scale opening with regard to the ventilation characteristics. The tendency of ventilation performance was the same as that of the basic opening. When a screen was attached to the window frame, C_d deceased to almost half that of the window without a screen.

3.6 Field Measurement for Cumulative Occurrence Frequency of Dimensionless Room Pressure

From the present wind tunnel experiment, the discharge coefficient was found to decrease rapidly in the P_R^* region of 0 to −5. In order to identify what level of P_R^* range should mainly exist during actual ventilation, field measurements were performed in a full-sized residential house, as shown in Fig. 3.13 (Ohba et al. 2004). In this house, the southern opening in an eastern room on the second

Table 3.2 Database of ventilation performance for various types of inflow openings

Opening	θ	C_{ds}	P_{RS}*	n	C_{ds} Ref. AIJ (1970)
Rectangular (longitudinal)		0.66	−12.21	0.15	0.65∼0.70
Rectanglar (lateral)		0.65	−12.88	0.20	
Square		0.64	−12.49	0.17	
Circular		0.61	−9.62	0.21	
Bell-mouth		0.84	−9.83	0.16	0.98
Oblong window louver (lateral opening)	90°	0.67	−9.78	0.18	0.70
	70°	0.65	−10.63	0.19	0.58
	50°	0.45	−9.53	0.11	0.42
	30°	0.29	−12.59	0.06	0.23
Window louver shorter side (longitudinal opening)	90°	0.62	−12.47	0.16	0.70
	70°	0.64	−17.43	0.15	0.58
	50°	0.58	−21.25	0.13	0.42
	30°	0.44	−33.16	0.07	0.23
Single pivoted window (longitudinal opening)	90°	0.61	−11.07	0.18	0.64
	70°	0.64	−15.78	0.18	–
	50°	0.57	−38.21	0.14	–
	30°	0.31	−14.52	0.08	–
Shorter side outrigger (longitudinal opening)	Outside 60°	0.59	−18.31	0.17	0.57
	Outside 30°	0.42	−24.31	0.17	0.38
	Inside 60°	0.60	−17.83	0.14	0.56
	Inside 30°	0.47	−9.37	0.12	0.38
Oblong outrigger window (lateral opening)	Outside 60°	0.56	−33.92	0.10	–
	Outside 30°	0.29	−193.22	0.04	–
	Inside 60°	0.56	−12.98	0.10	–
	Inside 30°	0.30	−416.32	0.02	–

floor had the highest ventilation flow rates among all the openings. The window comprised a 0.9 m^2 aluminum sash. Sliding shutters and eaves were located in the vicinity of the window frame.

Figure 3.14 indicates that the wind direction from south to southeast was prominent during the field measurement. Figure 3.15 shows the cumulative

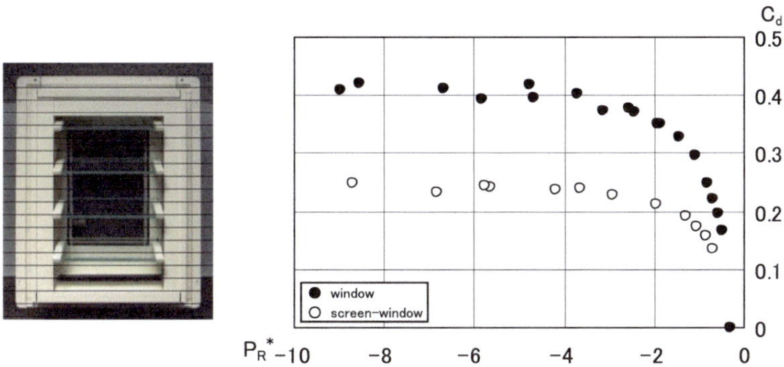

Fig. 3.12 Full-sized window frames and the discharge coefficient curves

Fig. 3.13 Full-sized residential house in field measurements

occurrence frequency of P_R^* for each mean wind direction. About 60 % of all the wind data were in the range of $|P_R^*| < 5$. This revealed that the ratio of the discharge coefficient decreased very frequently in actual winds. The discharge coefficient with the best fit with regard to wind direction must be clearly identified to obtain accurate ventilation airflow rates through the opening.

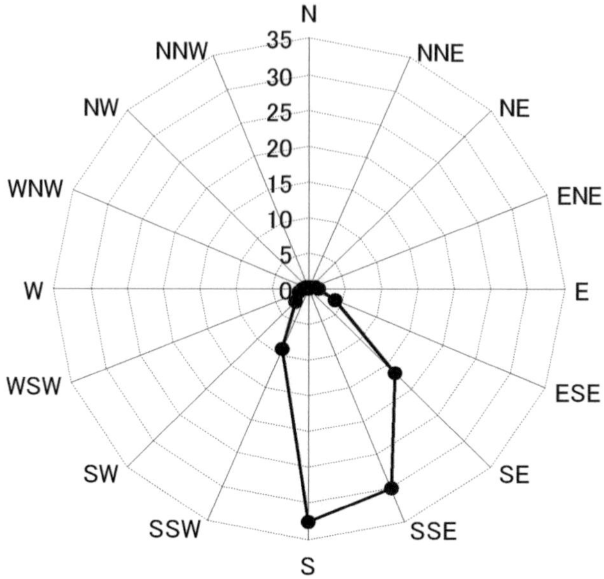

Fig. 3.14 Wind rose in field experiments

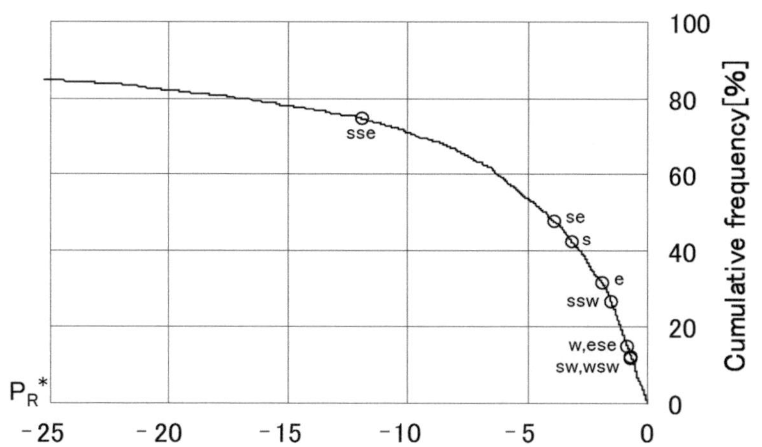

Fig. 3.15 Cumulative occurrence frequency distribution of room pressure in field experiments

3.7 Prediction Accuracy by Local Dynamic Similarity Model for Ventilation Flow Rates in Two Zones

3.7.1 Outline of Wind Tunnel Experiments

In order to validate the LDSM, wind tunnel experiments were conducted (Ohba et al. 2006). Figure 3.16 shows the two-zone room model, which had two rooms

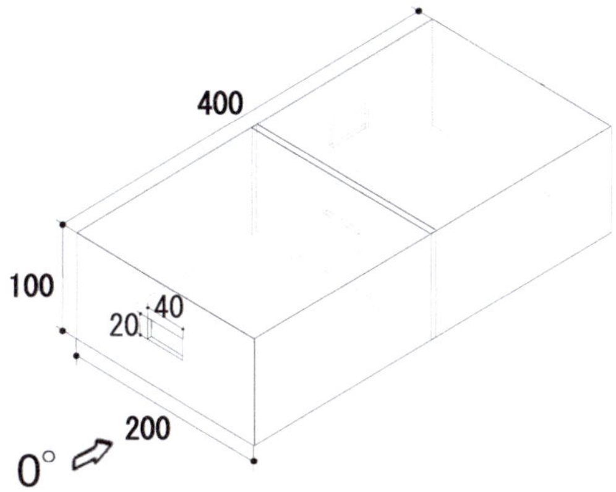

Fig. 3.16 Two-zone room model used in wind tunnel experiments

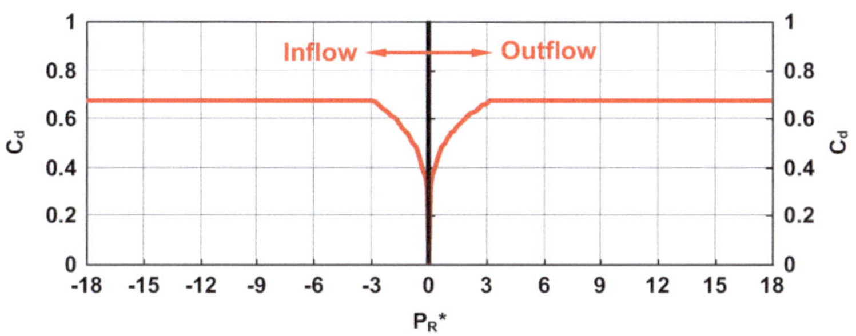

Fig. 3.17 Ventilation performance expressions for inflow and outflow openings

with openings 40 mm wide by 20 mm high. The opening in the partition wall was located at the center to reduce the effect of the inflow dynamic pressure on the outflow. The outflow opening was located at the side wall of the leeward room. The direction of the approach flow was varied in the range of −22.5° to 67.5°. The approach flow was a boundary layer flow with a power-law index of 0.25, and the reference velocity was kept at 7.0 m/s at the upwind edge of the model.

Figure 3.17 shows the ventilation performance for inflow/outflow openings (Goto et al. 2012). For a wind angle of −22.5°, the approach flow directly affected the airflow characteristics at the outflow opening so that the static pressure along the outflow greatly changed. For the outflow opening at a wind angle of −22.5°, different parameters from the prescribed ones were used. The C_{ds} to 0.79, P_{RS}* to 0.67, and n to 0.61 were set. The C_d of the opening in the partition wall was set to a constant value of 0.63.

3.7.2 Results of Prediction Accuracy of Discharge Coefficients and Ventilation Flow Rates

Figure 3.18 indicates the measured and predicted discharge coefficients for inflow and outflow openings (Tsukamoto et al. 2013). The experimental values of C_d were calculated from the measured ventilation flow rates, room pressures, and wind pressures. The conventional orifice model used a C_d of 0.67 for both inflow and outflow openings, regardless of wind angle. The experimental discharge coefficients were not fixed, but varied depending on wind direction. The LDSM had almost the same discharge coefficients as the experimental ones at the inflow opening. Even at the outflow opening, the predicted discharge coefficients followed the trend of the experimental discharge coefficients.

Figure 3.19 shows the prediction results of ventilation flow rates. The LDSM indicated better predictions for all wind angles than the conventional orifice model. The prediction accuracy increased by 5–20 % compared to the orifice model. The conventional orifice model cannot predict the ventilation flow rates accurately because the interfering cross-flow dynamic pressure P_t is too high at either inflow opening or outflow opening when the wind direction is not normal to the openings.

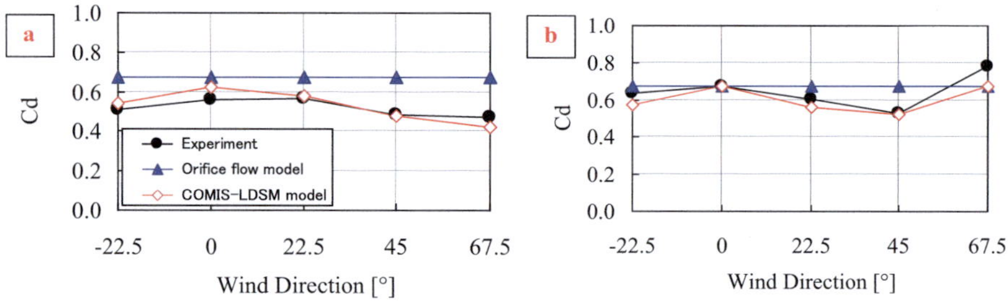

Fig. 3.18 Measured and predicted discharge coefficients: (**a**) inflow opening, (**b**) outflow opening

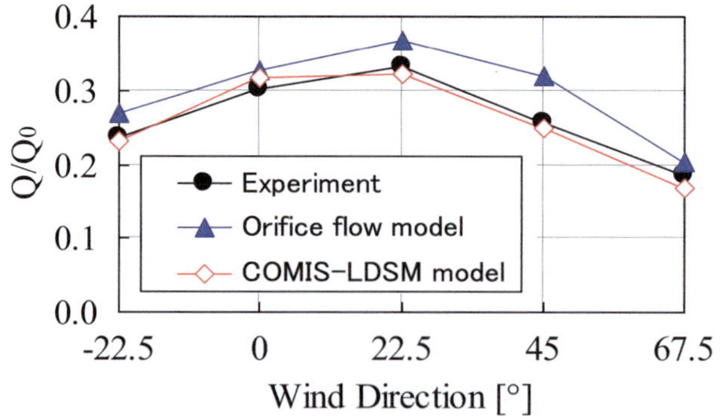

Fig. 3.19 Measured and predicted ventilation flow rates

3.8 Ventilation Network Model and Its Application

3.8.1 Ventilation Network Model

Figure 3.20 outlines the block diagram of the ventilation network model, which is composed of LDSM, COMIS model, and TRNSYS model (Tsukamoto et al. 2013). The COMIS code was revised on the basis of LDSM to calculate the discharge coefficients and airflow rates at inflow/outflow openings. P_W and P_t for the building envelope are provided as input data. The ventilation performances of inflow/outflow openings are also provided as input data. Arbitrary room pressure (P_R) is given as an initial condition, and a discharge coefficient corresponding to P_R* is selected from the ventilation performance curve. The calculation is performed by the relaxation-Newton method until ventilation flow rates of outflow and inflow in each room are balanced (Ohba et al. 2008). The energy simulation can be performed using TRNSYS code.

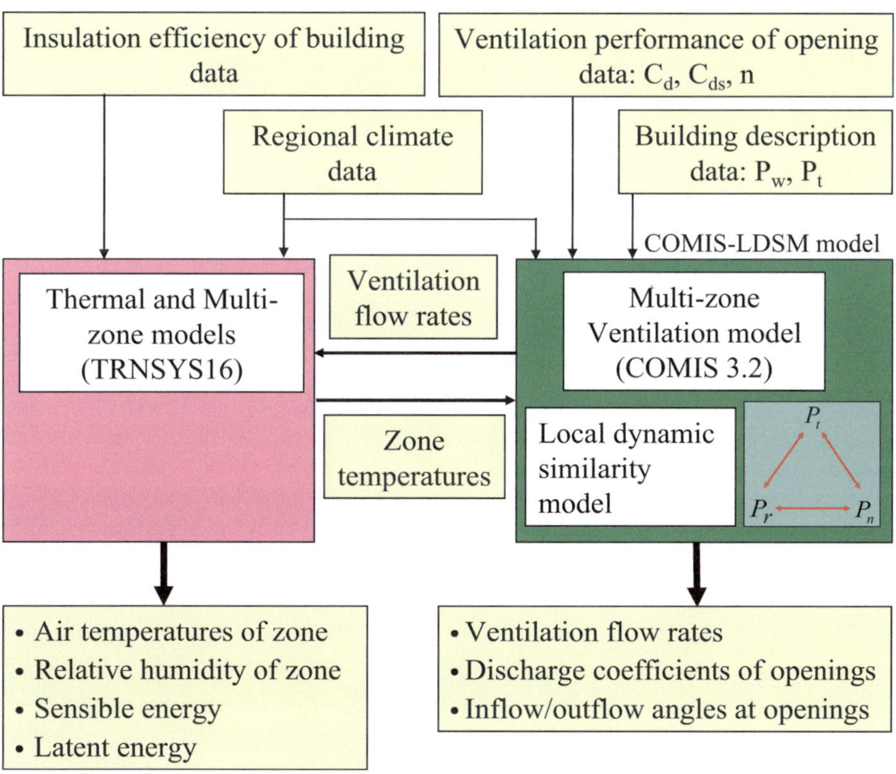

Fig. 3.20 Block diagram of ventilation network model composed of LDSM, COMIS model, and TRNSYS model

3.8.2 Detached House Model

A typically shaped house has been designated by the Architectural Institute of Japan (AIJ) (Udagawa 1985), as shown in Fig. 3.21. The family in the studied house was assumed to be composed of four persons: two children, a daily commuter, and a homemaker. It had 2 stories and a total floor area of 120 m^2, comprising six rooms including a living room kitchen. The eave height was 5.4 m. It was insulated according to the governmental next-generation energy standards.

Figure 3.22 indicates the layout of detached houses for gross coverage ratio of 20 %. The model scale was 1:100 in the wind tunnel experiment. In assuming the urban area, some dummy models were placed in a region from 2 m upwind to 0.5 m downwind of the target model.

The ventilation performance of inflow/outflow openings and roof opening was obtained through the data from Fig. 3.17 as the input data (Ohba et al. 2004; Kurabuchi et al. 2005a; Endo et al. 2009; Goto et al. 2012). The C_d of the internal doors was set to a constant value of 0.63 for the present calculation.

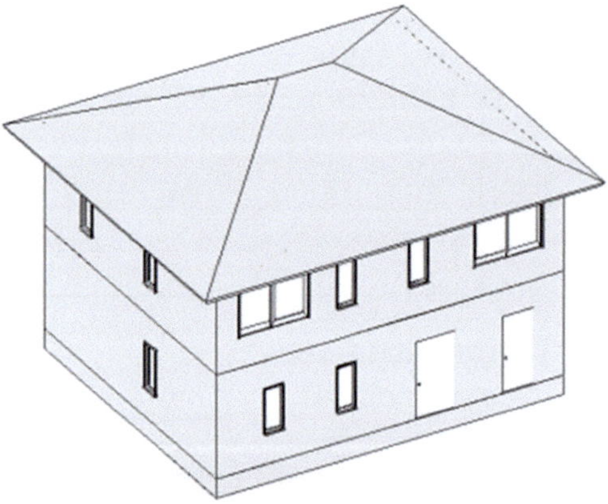

Fig. 3.21 AIJ detached house model

Fig. 3.22 Layout of detached houses for gross coverage ratio of 20 %

3.8.3 Simulation Study on Effects of Reducing Cooling Loads Through Cross-Ventilation in Detached House

3.8.3.1 Simulation Outline

The simulation study was performed for the AIJ detached house model to calculate the cooling load reduction effect when utilizing cross-ventilation (Ohba et al. 2008; Tsukamoto et al. 2009). Windows in habitable rooms and halls, and room doors were opened and closed according to the logic of opening/closing windows and the logic of turning on/off air-conditioners. Windows and doors in the toilets, bathroom, and washrooms were kept closed, as shown in Fig. 3.23. Expanded AMEDAS weather data (AIJ 2009) were used for the meteorological data in Tokyo, and the period was set to the month of June. The ventilation performance of the windows was shown in Fig. 3.17. The discharge coefficients of the room doors in the partition walls were set to a constant value of 0.63.

3.8.3.2 Logic of Opening/Closing Windows and Turning On/Off Air-Conditioner

The decision tree of operation of cooling and windows is shown in Fig. 3.24 (Nagai 2006). The logic of turning on/off air-conditioners was applied to only the LDK, the master bedroom, and the two children's rooms. Air-conditioners were not used in the other rooms. When the room temperature exceeded 28 °C in an occupied room, the air-conditioner was turned on, and when the room temperature dropped below

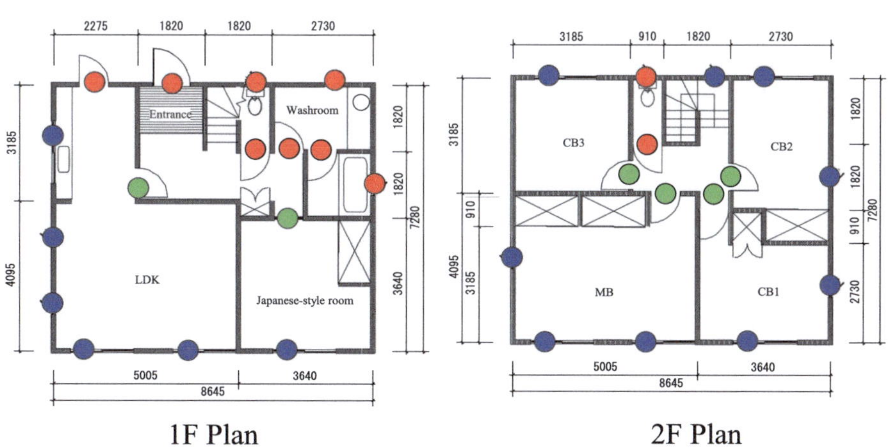

※ ● : Setting the logics of opening/closing windows and turning on/off air conditioning
※ ● : Opening doors only when windows are opened
※ ● : Windows and doors remained closed all the time

Fig. 3.23 Plan of AIJ residential model

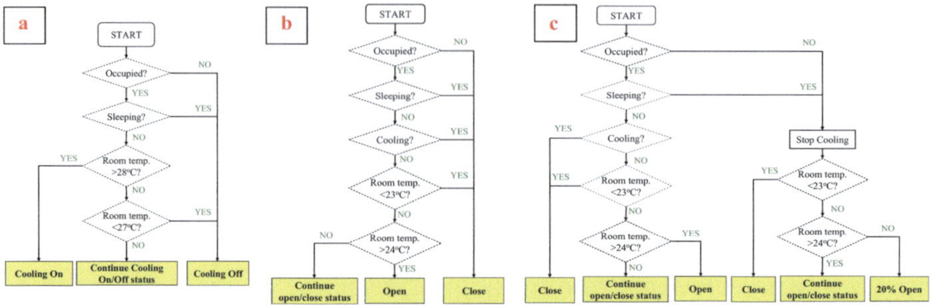

Fig. 3.24 Operation of cooling and window: (**a**) cooling, (**b**) basic window, (**c**) active window

Table 3.3 Schedule of occupied time zone

	1	2	3	4	5	6	7	8	9	10	11	12	13	14	15	16	17	18	19	20	21	22	23	24
LDK							1	4	1	1	1	1	1	1			3	3	3	3	2	2	1	
MB	2	2	2	2	2	2	1																1	2
CB1	1	1	1	1	1	1	1														1	1	1	1
CB2	1	1	1	1	1	1	1														1	1	1	1

27 °C, the air-conditioner was turned off. The cooling temperature controlled by the air-conditioners was set to 28 °C and the humidity to 60 %. The logic of opening/closing windows was applied to the windows and doors in the habitable rooms and the hall.

Three cases were simulated. For Case 1, we only used the logic of turning on/off air-conditioners with all windows closed. For Case 2, the basic logic of opening/closing windows was used. Windows were opened in occupied rooms when the temperature exceeded 24 °C and closed when the temperature dropped below 23 °C. Windows were closed when all family members were in sleep or when rooms were not occupied. For Case 3, the active logic of opening/closing doors was used. When the house was occupied (even when residents were asleep or the rooms were not occupied), we opened the windows 20 % when the room temperature exceeded 24 °C and closed them when the temperature dropped below 23 °C. Doors in habitable rooms were opened when the windows were opened, and the doors were closed when the air-conditioners were on and/or the windows were closed.

The schedule of persons at home was followed according to Table 3.3 proposed by AIJ (Udagawa 1985). The same pattern was used during weekdays and weekends. It was assumed that nobody occupied the house from 14:00 to 16:00 and the family was asleep from 23:00 to 6:00. It was also assumed that the Japanese room and a spare room were unoccupied at all time. Simulation was carried out every 15 min to evaluate window status and air-conditioner operation.

3.8.3.3 Simulation Results for Cumulative Cooling Loads in LDK for the Month of June

Figure 3.25 compares the duration of the operating time of the air-conditioners in June for gross coverage ratio of 0 %. Reduction of air-conditioner operating time of 98 h was achieved for the LDK and of 8–26 h for other habitable rooms by opening the windows during the occupied time zone in Case 2 versus Case 1. In Case 3, where the windows were opened liberally at night, further reduction of air-conditioner operating time of 120 h was achieved in the LDK and the air-conditioners did not need to be operated in the other habitable rooms.

Figure 3.26 compares the cumulative time the windows remained open for gross coverage ratio of 0 %. The cumulative time the windows remained fully open was longest in the LDK and shortest in the master bedroom in both Case 2 and Case 3. However, if the length of time the windows remained 20 % open is included, the length of time windows remained open in the master bedroom and the children's rooms exceeded 350 h in Case 3, over 120 h longer than in the LDK. Opening the windows 20 % greatly helped the utilization of cross-ventilation.

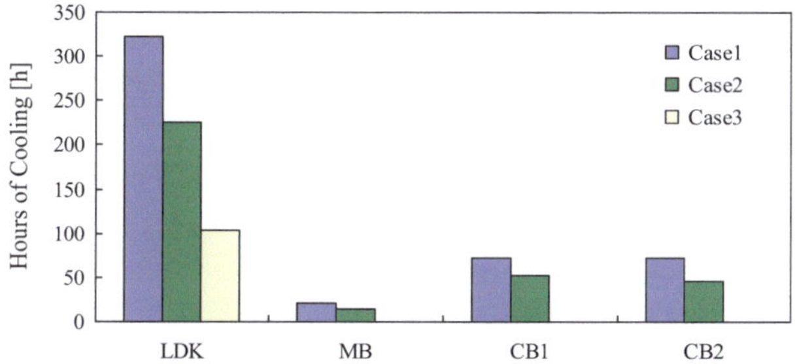

Fig. 3.25 Cumulative operating hours of air-conditioners in June

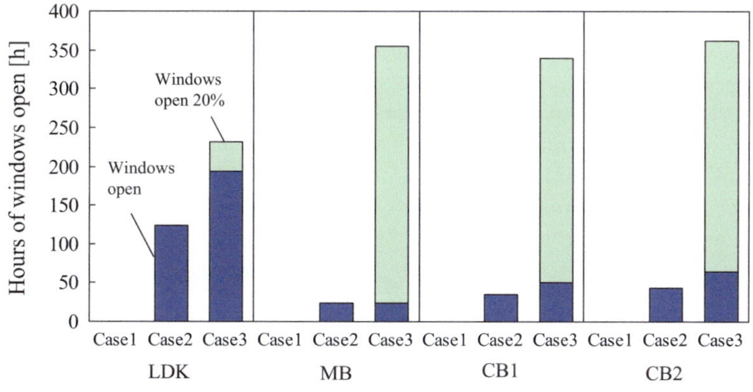

Fig. 3.26 Cumulative hours during which windows remained open in June

Table 3.4 Cumulative cooling load and reduction of cooling load in June for gross coverage ratio of 20 %

Case	Window operation	Cumulative cooling load [kWh]	Reduction of cooling load [%]
1	Closed	261	–
2	Basic opened/closed	247	5.1
3	Active opened/ closed	120	51.4

Table 3.4 shows the cumulative cooling loads in June and the effects of reducing energy through cross-ventilation in Cases 1, 2, and 3 for the building coverage ratio of 20 %. Utilization of cross-ventilation succeeded in reducing cooling loads by only 5 % compared to those required when the windows were closed. However, when the windows remained open during the unoccupied time zone or while residents were sleeping, the cooling load was reduced by 51 % compared to that when all windows remained closed.

3.9 Concluding Remarks

This chapter has described the ventilation flow structure and introduced a new ventilation network model. These can be summarized as follows:

- The ventilation air flow after flowing into the room turns steeply downward due to the transport of downward momentum immediately before the opening and the pressure gradient caused by the circulating flow developed at the lower portion of the building.
- Before the ventilation air flow reaches the opening, the total pressure of the approach flow is preserved almost completely regardless of the approaching flow angle if flow separation does not occur before it reaches the opening.
- The pressure loss at the opening depends on the contraction of ventilation air flow and on static pressure loss consumed for the production of turbulent kinetic energy.
- In most cases, ventilation air flow passes along the front surface of a building immediately before reaching the opening, and it has an air flow component tangential to the opening.
- The local dynamic similarity model (LDSM) is capable to predict the ventilation flow rate with better accuracy. LDSM has superiority over the conventional orifice flow model when discharge coefficients decreased due to changes of wind directions.
- The LDSM provided us with valuable information on inflow/outflow angles at openings. It is very useful for predicting flow patterns in rooms. The prediction

accuracy of ventilation flow rates was increased by 5–20 % compared with the conventional orifice model that uses fixed discharge coefficients.
- In field measurements of a full-sized residential house, about 60 % of the measured wind data fell into the range of $|PR^*| < 5$. This revealed that the decrease ratio of discharge coefficient was very high in actual winds.
- A ventilation network model based on local dynamic similarity theory was coupled with COMIS model and TRNSYS model, which are widely used for energy conservation simulation. A simulation study was performed on cooling loads in a detached house. A reduction in cooling load of around 50 % could be achieved when the windows were opened liberally at night compared with the cooling load when all the windows remained closed for the building coverage ratio of 20 %.

Acknowledgement The author indebted to the contribution of his former students and post-doctoral fellows on the subject of this paper.

References

Architectural Institute of Japan (1970) Handbook of ventilation design, vol 18. Shokokusha, Japan
Architectural Institute of Japan (2009) Expanded AMEDAS weather data 1981–2000
Aynsley RM (1988) A resistance approach to estimating airflow through building with large openings due to wind. ASHRAE Trans 94(Part 2):1661–1669
Endo T, Kurabuchi T, Akamine Y, Ohba M, Kamata M (2009) Development of the test procedure and the database of cross-ventilation characteristics of various openings. J Environ Eng AIJ 74 (646):1315–1320
Goto T, Ohba M, Kurabuchi T et al (2012) Investigation of the application of local dynamic similarity model to outflow openings prediction accuracy of flow rate of cross-ventilated buildings (Part 3). J Environ Eng AIJ 77(674):259–266
Heiselberg P (2002) Principles of hybrid ventilation. Hybrid Ventilation Centre, Aalborg University, Denmark
Ishihara M (1969) Building ventilation design. Asakura Shoten, Japan
Kurabuchi T, Ohba M, Arashiguchi A, Iwabuchi T (2000) Numerical study of airflow structure of a cross-ventilated model building. In: Proceedings of ROOMVENT 2000, UK, pp 313–318
Kurabuchi T, Ohba M, Iwabuchi T (2002a) Verification of simulated results and streamtube analysis for normal flow angle case analysis of airflow structure of a cross-ventilated building based on large eddy simulation and wind experiment (Part 1). Environ Eng A 561:47–52
Kurabuchi T, Ohba M, Fugo Y, Endo T (2002b) Local similarity model of cross-ventilation: part 1 modelling and validation. In: The 8th international conference on air distribution in rooms 'ROOMVENT 2002', Denmark, pp 613–616
Kurabuchi T, Ohba M, Endo T, Akamine Y, Nakayama F (2004) Local dynamic similarity of cross-ventilation, part 1: theoretical framework. Int J Vent 2(4):371–382
Kurabuchi T, Ohba M, Goto T, Akamine Y, Endo T, Kamata M (2005a) Local dynamic similarity concept as applied to the evaluation of discharge coefficients of cross ventilated buildings, Part 1 Basic idea and underlying wind tunnel tests; Part 2 Applicability of local dynamic similarity model; Part 3 Simplified method for estimating dynamic pressure tangential to openings of cross-ventilated buildings. Int J Vent 4(3):285–300
Kurabuchi T, Ohba M, Endo T (2005b) Verification and streamtube analysis of simulated results of airflow of a cross-ventilated buildings for various wind incident angles analysis of airflow of

cross-ventilated buildings based on LES and wind tunnel experiment (Part 2). J Environ Eng AIJ 591:7–13
Murakami S, Mochida A, Kondo K, Ishida Y, Tsuchiya M (1996) Development of new k-ε model for flow and pressure fields around bluff body. CWE96, Colorado
Nagai T (2006) Windows and HVAC operation to reduce cooling requirement by means of cross-ventilation. Int J Vent 5(1):151–162
Nonaka T, Kurabuchi T, Ohba M et al (2009) Wind tunnel experiments on surface wind pressure and cross-ventilation flow rate in densely populated residential area study on proper design method of locating windows aiming at utilization of cross-ventilation in densely populated residential area (Part 1). Environ Eng AIJ 74(642):951–956
Ohba M, Kurabuchi T, Fugo Y, Endo T (2002) Local similarity model of cross-ventilation Part 2 Application. In: The 8th international conference on air distribution in rooms 'ROOMVENT 2002', Denmark, pp 617–620
Ohba M, Kurabuchi T et al (2004) Local dynamic similarity of cross-ventilation, Part 2: application of local dynamic similarity model. Int J Vent 2(4):383–393
Ohba M, Goto T, Kurabuchi T et al (2006) Experimental study on predicting wind-driven cross-ventilation flow rates and discharge coefficients based on the local dynamic similarity model. Int J Vent 5(1):105–114
Ohba M, Kurabuchi T, Goto T et al (2008) Study on prediction of ventilation flow rates in detached house based on coupled simulation of semi-empirical envelope flow model and network model. in: Proceedings of 4th international conference on advances in wind and structures, Korea, pp 1156–1166
Sandberg M (2002) Airflow through large openings – a catchment problem? In: Proceedings of ROOMVENT2002, Denmark, pp 541–544
Sawachi T (2002) Detailed observation of cross ventilation and airflow through large openings by full scale building model in wind tunnel. In: The 8th international conference on air distribution in rooms 'ROOMVENT2002', Denmark, pp 565–568
Shohda T, Sekine T (1961) Experimental study on ventilation through building openings Part1 – on pressure loss coefficient and wind pressure coefficient of openings. Trans Archit Ins Jpn 68, pp 116–120
Tsukamoto K, Ohba M, Kurabuchi T, Nonaka T, Goto T (2009) Study on reduction of cooling loads in detached house by cross-ventilation using coupled simulation of semi-empirical ventilation model and network models. ROOMVENT2009, Korea, e-format
Tsukamoto K, Ohba M, Kurabuchi T et al (2013) Prediction accuracy of ventilation flow rates in multiple interzones by using local similarity model and application of the model to analysis of ventilation flow rates in detached house prediction accuracy of flow rate of cross-ventilated buildings (Part 4). J Environ Eng AIJ 78(684):157–163
Udagawa H (1985) Typical problems for house. In: 15th heat symposium, Japan
Vickery BJ, Karakatsanis C (1987) Experimental wind pressure distribution and induced internal ventilation flow in low-rise industrial and domestic structures. ASHRAE Trans 93(Part 2): 2198–2213

Chapter 4
Passive Cooling of Buildings: Present and Future Needs: Recent Progress on Passive Cooling Convective Technologies

Mattheos Santamouris and Denia Kolokotsa

Abstract The maximisation of energy efficiency in buildings can be performed by successful demonstration of the passive technologies. Passive cooling is passing to a phase of maturity, while significant research has been performed worldwide. The aim of this chapter is to underline and review the recent state-of-the-art technologies for passive cooling convective strategies for buildings and their contribution in the improvement of the indoor environmental quality as well as in the reduction of cooling needs. The paper starts with a short introduction in passive cooling needs and continues with the analysis of the two passive cooling convective techniques, i.e. ground cooling and night ventilation.

Keywords Passive cooling • Ground cooling • Ventilative cooling

4.1 Introduction

The building sector represents the most important consumer of energy with an average share close to 40 %. While in the developed world heating seems to be the more consuming of the specific building energy needs, cooling is a very intensive energy activity presenting a continuously increasing share in the overall energy budget of the buildings sector. In fact, the energy consumption for cooling purposes increases almost everywhere in the world because of the improved standards of life, the relative affordability of air-conditioning systems in the developed world, the inappropriateness of buildings to avoid indoor overheating due to the adoption of design standards proposed by a rather universal modern architecture, and, the most important, because of the serious temperature increase in the urban environments (Antinucci et al. 1992).

M. Santamouris (✉)
Physics Department, University of Athens, Panpeistimioupolis, Athens, Greece
e-mail: msantam@phys.uoa.gr

D. Kolokotsa
Technical University of Crete, Chania 731 00, Greece

Y. Tamura, R. Yoshie (eds.), *Advanced Environmental Wind Engineering*,
DOI 10.1007/978-4-431-55912-2_4

Global climatic change and heat island increase urban temperatures and extreme heat-related climatic phenomena. The ambient temperature of numerous urban locations has increased considerably in the last years because of the heat island phenomenon. Heat island relies on higher urban temperatures compared to the suburban or rural ones because of the positive thermal balance of cities. Important research carried has permitted to fully document the intensity of heat island in Europe and the other continents (Santamouris 2007).

Several studies have shown the relation between heat island and the excess need for cooling. Higher urban temperatures increase the absolute energy consumption of the buildings during the summer period and raise the peak electricity demand (Akbari et al. 1992; Cartalis et al. 2001; Hassid et al. 2000; Santamouris et al. 2001). In parallel, the heat island phenomenon intensifies the stress in urban areas (Pantavou et al. 2011), increases pollution in cities as higher urban temperatures, accelerates the rate of photochemical ozone production (Stathopoulou et al. 2008), and expands the ecological footprint of the cities (Santamouris et al. 2007b).

Apart from the heat island phenomenon, the global climatic change observed not only in urban areas has a serious impact on the energy consumption of buildings for cooling purposes. Cooling degree days are considerably increased during the last years in most zones of the planet, while the future cooling energy demand seems to be much higher than some decades ago. For example, an analysis of 40 years of hourly data series from nine meteorological stations in Greece performed to understand the impact of air temperature, and relative humidity trends on the energy consumption of buildings has shown that for the period in question, the heating load in the Greek building sector has decreased by about 1 kWh/m^2 per decade, while the cooling load increased by about 5 kWh/m^2 per decade (Kapsomenakis et al. 2013). In parallel, many studies examined the relation of future temperatures with the cooling energy needs of buildings. All studies have shown that very important energy consumption has to be expected unless severe technological measures are taken. In particular, recent research has shown that because of the expected serious future temperature increase, the heating energy demand of the building sector in Greece could decrease by about 50 %, while the corresponding energy consumption for cooling could increase by 248 % until 2100 (Asimakopoulos et al. 2012).

High ambient temperatures have a serious impact on the quality of life of the general population but mainly on the so-called low-income or vulnerable population. It is well known that low-income population has to spend more energy than high-income people, to satisfy the specific cooling and heating demands per square metre because of the inferior quality of the building envelope and lower energy performance of the houses they are living in. It is characteristic that the low-income population in Greece has to spend almost 120 % more per square metre and inhabitant than the high-income group to satisfy the specific cooling needs of the houses (Santamouris et al. 2007a).

The problem is quite dramatic for the vulnerable and low-income population that cannot afford to pay for air-conditioning. Measurements performed in Athens during the heat waves of 2007 in some tenths of low-income houses have shown that the inhabitants lived in temperatures above 34 °C for more than 200 continuous hours (Sakka et al. 2012). The specific indoor conditions put a serious threat to the health of low-income citizens and increase highly the mortality rates.

Passive cooling systems and in particular technologies based on convective heat amortisation systems have achieved a very high degree of development (Santamouris and Kolokotsa 2013). The new proposed technologies have proven very efficient and when used may improve the indoor environmental quality and decrease the energy cost of buildings. As mentioned by Santamouris et al. (2007c), the new proposed and developed technologies are characterised by low cost and are easy to apply.

The present article investigates the present and future needs for cooling while it tries to present the more promising new developments on the field of convective heat amortisation techniques.

4.2 Present and Future Needs for Cooling

As mentioned in the previous sections, the important climate change and heat island phenomenon have increased considerably the cooling load of buildings. Several studies are performed to investigate the impact of heat island on the cooling needs of buildings.

Studies in the USA have shown that in cities with population larger than 100,000 citizens, the electricity load increases by 1.5–2 % for every degree F of temperature increase (Akbari et al. 1992). Given that the temperature increase in some US cities during the summer afternoons is between 2 and 4 °F, it is calculated that 3–8 % of the current demand of electricity in the urban areas is used to compensate the heat island phenomenon. In parallel, in Los Angeles, USA, it is found that the net rate of increase of the electricity demand is close to 300 MW per degree F. It is also estimated that the absolute temperature increase in the city since 1940 is 5 F, which is translated to an additional electricity demand of 1.5 GW attributed fully to the heat island phenomenon. Akbari has calculated that in the USA the electricity costs to compensate heat islands could be as much as 1 million dollars per hour or almost 1 billion dollars per year (Akbari et al. 1992).

In Athens, Greece, the spatial distribution of the cooling load of a typical building is calculated using temperature data from a very high number of temperature stations. Figure 4.1 gives the iso-cooling lines in kWh/m^2 and month. As shown, the necessary cooling load at the centre of the city is almost double than in the suburban areas. Much higher values are calculated in zones of high density and high anthropogenic heat generated (Santamouris et al. 2001).

Apart from increased energy loads for cooling, high ambient temperatures increase the peak electricity loads and put a serious strength on the local utilities. It is calculated that the peak cooling demand in Athens, present a very high increase because of the heat island phenomenon. In particular, while the peak demand for a typical office in the suburban zones of Athens is close to 13.7 kW, the corresponding load at the centre of the city is 27.5 kW.

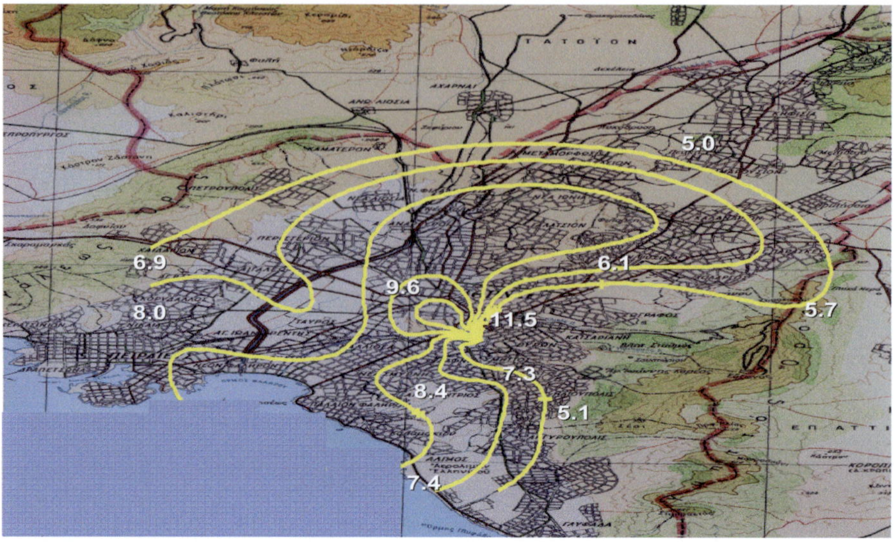

Fig. 4.1 Distribution of the cooling load in the city of Athens for a set point temperature of 27 °C and for August 1996

The impact of climate change on the ambient temperature is already very important, and it is expected to further increase if additional mitigation and adaptation measures are not taken. Forty years of ambient temperature have been used to calculate the evolution of cooling and heating loads in typical buildings for various cities in Greece (Kapsomenakis et al. 2013). Data covered the period 1970–2010. The calculated increase of the cooling load is shown for four major cities in Fig. 4.2. As shown, the cooling energy demands have increased seriously in the last 40 years, and the corresponding increase rate is close to 5 kWh/m^2 per decade. A further analysis for the same zone of Europe has attempted to identify the future energy needs of buildings considering the various scenarios proposed by IPCC regarding the evolution of the ambient temperature (Asimakopoulos et al. 2012). It is found that the expected increase of the cooling needs for a residential building considering the A1B scenario for 2041–2050 will be 83 % compared to the actual situation and 167 % for the period 2091–2100. The results are given in Fig. 4.3 for three types of residential buildings: a building constructed using technology of 1980, a contemporary building and a passive house. It is evident that because of the important temperature increase, cooling loads will skyrocket unless appropriate mitigation and adaptation techniques are considered.

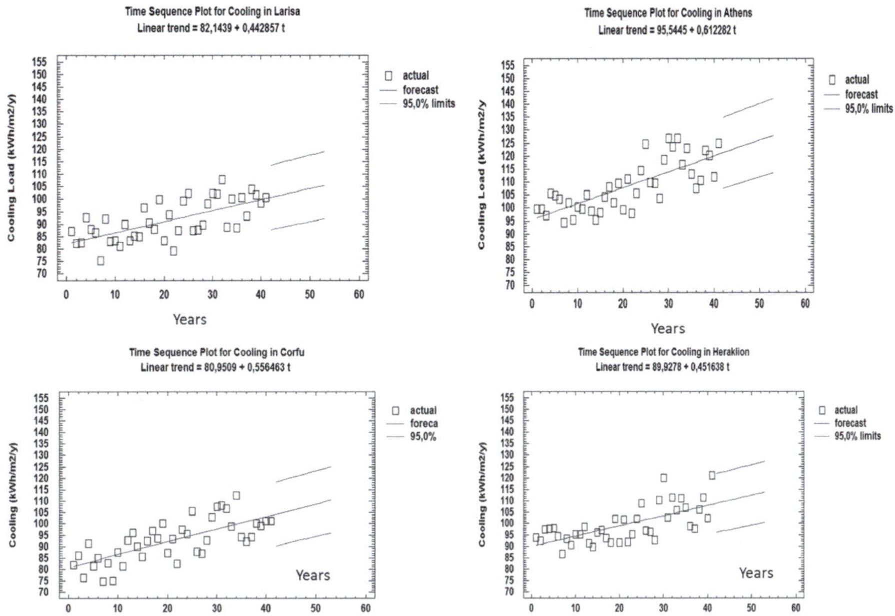

Fig. 4.2 Annual variation of the cooling load in four major Greek cities (Kapsomenakis et al. 2013)

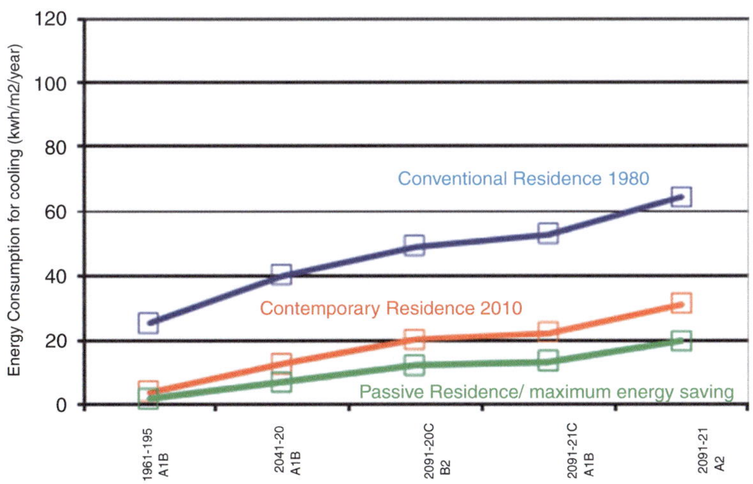

Fig. 4.3 Change of the energy consumption for cooling of a conventional (*black*), a contemporary (*red*) and a passive (*green*) residence in Attica for all climatic scenarios (Asimakopoulos et al. 2012)

4.3 Passive Cooling to Fight Vulnerability and Protect Low-Income Population

High urban temperatures have a very serious impact on the health of urban population. High ambient temperatures are found to cause important cardiovascular, respiratory and cerebrovascular disorders (McMichael et al. 1996). In parallel, high ambient temperatures decrease the viscosity of the blood and increase the risk of thrombosis, while older residents have problems of thermoregulation and impaired kidney function. Various studies investigating the relation of the ambient temperature against hospital admissions have shown that during warm periods, the number of admissions increases for conventional heat-related diseases like heat stroke, heat exhaustion, neurological conditions mental illness and renal diseases (Kovats et al. 2004).

Exposure of human beings in extreme high temperatures may result in increased human losses. Analysis of the impact of the last decade heat waves has shown that mortality increases highly, while most of the victims belong to the low-income and vulnerable groups. Baccini et al. (2008) examined the impact of high temperatures on excess mortality for 15 European cities and reported that the threshold for Mediterranean cities is close to 29.4 °C while it is much lower 23.3 °C for Northern and Condimental cities. As reported, increase of the temperature by 1 C above the threshold increases mortality by 3.12 % in the Mediterranean and 1.84 % in Northern Europe. Figure 4.4 reports in a comparative way the transfer functions

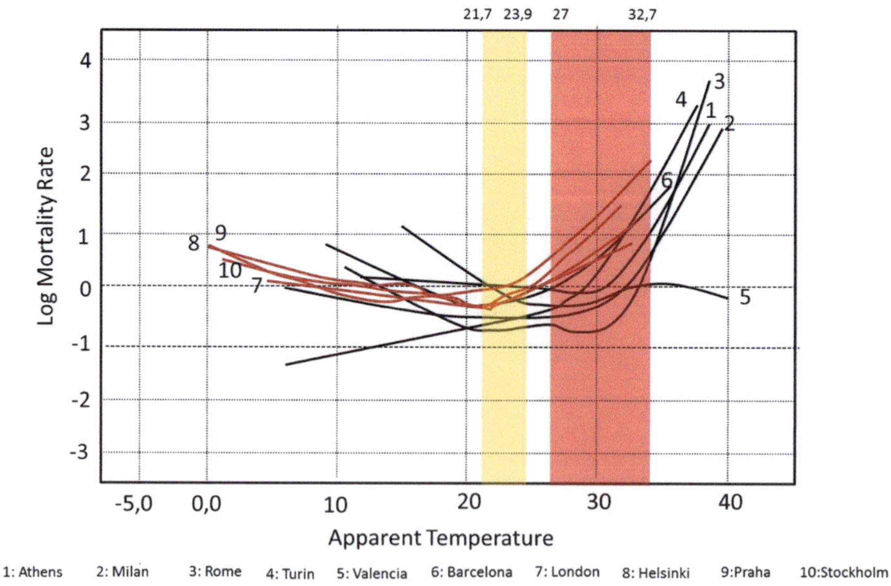

Fig. 4.4 Transfer functions between ambient temperature and the Log of Mortality Rate as reported by Baccini et al. (2008). Also, the specific bands for the Mediterranean and Northern countries where mortality starts to rise are given

between the ambient temperature and the log of the mortality for some Mediterranean and Northern cities as reported by Baccini et al. (2008).

The use of advanced passive cooling techniques may provide a very reliable solution to avoid health problems and decrease mortality during heat waves. The use of heat dissipation techniques like natural and nighttime as well as the use of low-cost evaporative and ground systems may improve the environmental quality of vulnerable population and save lives.

4.4 Recent Progress on Convective Passive Cooling Techniques

4.4.1 Ventilative Passive Cooling

It is widely known that the oldest, straightforward convective passive cooling method admits cool night air to drive out the warm air. Nocturnal convective cooling exploits the cold night air to cool down the building's absorbed heat gains during daytime and reduce the daytime temperature rise. Night ventilation can either be driven by natural forces – i.e. stack or wind pressure difference – or may be sometimes supported by a small fan power to provide sufficient airflow at times when the natural forces are weak. As a consequence, temperature peaks are reduced or even postponed. For specific climatic conditions, the cooling potential of night ventilation techniques depends on the air flow rate, the thermal capacity of the building, and the appropriate coupling of the thermal mass and the air flow. As a result, the effectiveness of night ventilation techniques is determined by the prevailing climatic conditions, the microclimate, the building characteristics and the location (Geros et al. 2005; Santamouris et al. 1996). The outdoor temperature, the relative humidity and the wind speed are the environmental parameters that influence the successful application of night ventilation techniques. Other parameters (Kolokotroni and Aronis 1999) can be the building mass, glazing ratio, solar and internal gains and orientation.

To better understand the relative phenomena and also quantify the impact of convective nocturnal ventilation techniques, important experimental and theoretical research has been carried out (Santamouris et al. 2010).

Santamouris et al. (2010) pointed out that the application of night ventilation techniques to residential buildings may lead to a decrease of cooling loads almost 40 kWh/m^2/y with an average contribution of 12 kWh/m^2/y. In urban areas though, the urban heat island (UHI) phenomenon deteriorates quality of life and has a direct impact on the energy demand, the environmental conditions and, consequently, ventilation effectiveness. The increased urban temperatures (Livada et al. 2002, 2007) exacerbate the cooling load of buildings, increase the peak electricity demand for cooling, decrease the efficiency of air-conditioners (Cartalis et al. 2001; Santamouris et al. 2001) and create an emerge necessity for passive cooling.

Specific monitoring studies in real buildings or test cells are reported in Allard and Santamouris (1998), Blondeau et al. (1997), Geros et al. (1999, 2005), Givoni (1994), Kolokotroni et al. (1997), Krüger et al. (2010), Kubota et al. (2009), Pfafferott et al. (2003), Santamouris and Assimakopoulos (1997), Springer et al. (2005), van der Maas and Roulet (1991) and Zimmerman and Andersson (1998). Most of the studies conclude that the use of convective ventilation in free-floating buildings may decrease the next day peak indoor temperature up to 3 K. In parallel, when applied in air-conditioned buildings, a considerable reduction of the peak cooling may be expected.

Available design methods for night ventilation range from simple analytical and empirical methods to multi-zone and computational fluid dynamics (CFD) techniques (Jiru and Bitsuamlak 2010; Li and Nielsen 2011; Li and Heiselberg 2002).

Most analytical approaches are based on a conventional macroscopic approach that utilises the Bernoulli equation for flow and opening (Allard and Santamouris 1998). This equation, which is based on the conservation of energy, is used to calculate air velocities through openings.

Network models are widely used for simulation of ventilation in multi-zone buildings coupled with thermal flow analysis (Li and Heiselberg 2002). The network method is based on the application of the Bernoulli equation to determine the pressure difference and hence flow rate across each opening in the flow network (Flourentzou et al. 1998). Zones are interconnected by flow paths, such as cracks, windows, doors and shafts, to form a flow network. Network methods are able to take into account the effects of outdoor climate, the location and size of each opening, and stack, wind and mechanically driven ventilation.

Computational fluid dynamics are based on the Navier-Stokes equations. CFD simulation provides detailed distribution of air temperature, air velocity, contaminant concentration within the building and its surrounding areas. However, the application of CFD for convective cooling prediction has been limited due to increased computational time and computer requirements. Building simulation tools facilitate energy-efficient, sustainable building design by providing rapid prediction of thermal comfort, indoor air flow of the building and better understanding of the consequences of ventilation for cooling. Another modelling software that is very frequently used is TRNSYS coupled with COMIS (Duffy et al. 2009; Fiksel et al. 1995). The specific technique is a combination of thermal modelling with air flow representation and is a very powerful tool for estimating thermal comfort coupled with indoor air quality (Fraisse et al. 2010; Manz and Frank 2005).

Moreover convective cooling via ventilation is a suitable technique for various building types including offices, residential, industrial, etc. Some examples are included below:

- Concerning office buildings, a significant number of studies are showing the ventilative cooling effectiveness. Blondeau et al. (1997) studied the night ventilation in offices during summer period, showing a reduction of diurnal variation from 1.5 to 2 K. Kolokotroni et al. (Kolokotroni et al. 1998; Kolokotroni 2001) showed that the energy savings for night ventilation in UK offices can be about 5 %.

- For residential buildings, various studies can be found. For example, the effectiveness of night ventilation technique for residential buildings in hot-humid climate of Malaysia is analysed in Kubota et al. (2009). The effects of different natural ventilation strategies on indoor thermal environment for Malaysian terraced houses are evaluated based on the results of a full-scale field experiment. The results indicated that the cooling effect of night ventilation is larger than those of the other ventilation strategies during the day and night. Night ventilation with solar chimneys is applied to social housing design for hot climates in Madrid (Macias et al. 2006). Storage chimneys – oriented to the west – collect solar gains during the afternoon, while the surface temperatures of the concrete walls reach temperatures up to 50 °C. The fresh cold night air enters through the east facade and runs through the flat, cooling down the thermal masses of the open walls and ceilings. Results of the simulation show that indoor temperature remains between 21 °C and 23 °C during the night. Two hundred and fourteen air-conditioned residential buildings using night ventilation techniques have been analysed in Santamouris et al. (2010). It has been found that night ventilation applied to residential buildings may decrease the cooling load up to 40 kWh/m^2 per year with an average contribution close to twelve, 12 kWh/m^2 per year .
- Night ventilation coupled with double skin facades for industrial archaeology buildings is studied in Ballestini et al. (2005) showing using simulation techniques that at least 12 % of energy can be reduced on yearly basis.
- A night-ventilated library located in Ireland is considered and analysed in Finn et al. (2007). The building is modelled using ESP-r, and the mean bias deviation between the predicted and experimental data is better than 0.45 °C for the dry bulb temperature. Examination of night ventilation rates indicated that increasing night ventilation up to 10 air changes per hour result to 1 °C reduction for medium gains and 2 °C reduction for high internal gains.

Therefore, night ventilative cooling is a very effective method to reduce the air-conditioning demand for any building type and improve thermal comfort during daytime regardless the climatic conditions.

Night ventilation is one of the most cost-effective and efficient passive cooling techniques for low-income households (Springer et al. 2005). Studies show that in hot climates, 12 ach/h during night with 1 ach during the day may provide comfortable indoor conditions (Golneshan and Yaghoubi 1990). A careful design of opening is very important for the improvement of the night ventilation effectiveness in urban regions (Geros et al. 1999; Kolokotroni 2001; Maragogiannis et al. 2011; Kolokotroni et al. 2006).

4.4.2 Ground Convective Cooling

Another convective cooling strategy is the drawing of outdoor air through tubes buried in the ground and dumped into the dwelling. Made of material that allows easy thermal transfer, these tubes are buried a few metres deep to avoid the warmer daytime surface temperatures. The designation of this system varies, as it could be referred to as "earth tubes" or as "ground-coupled air heat exchangers".

Many applications of ground convective cooling are available around the world. Ground convective cooling has been applied in various types of buildings as listed below:

- The application of ground convective cooling in an office building in Belgium is described in Breesch et al. (2005). The building has a surface around 2000 m^2. The used earth-to-air heat exchanger includes two concrete pipes of 80 cm internal diameter and 40 m length, buried at depths 3 and 5 m respectively. Monitoring of the building has shown that the maximum supply of air from the buried pipes never exceeds 22 °C. In parallel, it is found that ground convective cooling decreases the discomfort hours in the office by 20–30 % during the whole summer period.
- The application of ground convective cooling in three office buildings in Germany is described in Pfafferott et al. (2003). During the summer period when the ambient temperature was above 30 °C, the exit temperature in the exchanger was close to 18 C. The outlet air temperature was always between 20 °C and 5 °C. The specific energy gains were calculated close to 13.5 Kwh/m^2 per year.
- Two polyethylene pipes having a length of 90 m and buried at 2.80 m depth are used to supply cooling to an office building of 1488 m^2 in Germany. The system is monitored for about 3 years, and it is found that the temperature drop in the pipes was close to 10 °C, while the COP of the system was measured between 35 and 50.
- The application of almost 28 buried pipes in a school in Italy is described in Grosso and Raimondo (2008). The pipes' length was 70 m and was buried at 2.6 m. The building was not monitored, but simulation results show that the average cooling contribution of each pipe was close to 760 kWh/m^2 per year.
- Four ground pipes buried at 1.5 m depth are used to provide cooling to the philosophical school of the University of Ioannina, Greece. The building has a surface of 4100 m^2. The system was monitored and is found that it provides 33 kWh/m^2 per year for cooling purposes.
- The application of an earth-to-air heat exchanger to a hospital building in India is described in Badescu and Isvoranu (2011). The monitoring of the system showed that when the maximum ambient temperature was close to 42 °C, the maximum air temperature at the exit of the tubes was close to 25.5 °C
- A ground cooling system composed by buried pipes of 40 m length, buried at 4 m depth, is installed in a building named "CircoLab", which is a multifunctional facility with corporate, conference, recreational, cultural and community functionalities. The surface of the building was 382 m^2. Monitoring

of the system has shown that the exit temperature from the buried pipes was almost 5 °C lower than the ambient temperature.

- The use of ground convective cooling in a multi-storey passive house building in Romania, (AMVIC PH), is reported in Badescu and Isvoranu (2011). The ground exchangers consisted of two parallel drums of external diameter 400 mm buried at 3.5 m depth. Monitoring showed that when the ambient temperature was 32 ° C, the temperature at the exit of the pipes was close to 26 °C.
- The application of an earth-to-air heat exchanger in a passive house in France is described in Thiers and Peuportier (2008). The pipes are buried at 1.6 m and have a length of 30 m. The system is connected to a building of 132 m^2 and was used for ventilation and cooling purposes. The project is not monitored, but simulation results show that the use of the EATHE reduces the cooling degree days from 56–22. Another application of ETAHE in residential buildings in France is described in Trombe et al. (1991). It is found that when the ambient temperature was 37 °C, the exit temperature from the exchangers was 24 °C. In parallel, the building connected to the EATHE was almost 3 °C lower temperature than the same building not associated to an exchanger.

In addition ground convective cooling for cities is proposed by various researchers. Fintikakis et al. (2011) evaluated the microclimatic modifications that were applied in the historic centre of Tirana, Albania. Among other mitigation techniques, ground convective cooling was applied. The researchers found that the maximum temperature drop due to all the applied techniques was 3 K, while the maximum contribution of the earth-to-air heat exchangers was found to be close to 0.7 K.

Ground cooling and the earth-to-air heat exchangers can be also a viable solution for low-income households. A significant number of buildings have been designed and monitored, and the performance of the system has been proven very high (Thiers and Peuportier 2008). For example, the use of earth-to-air heat exchangers in a new housing development in Portugal has contributed to a reduction in the cooling needs by 95 % compared to an air-conditioned building, though the mean cost per building was quite low, at close to 7500 €/house (Santamouris and Kolokotsa 2013).

4.5 Conclusions

In this chapter, an analysis of the convective passive cooling techniques for buildings is performed. Moreover, the applicability of those techniques in low-income households is revealed. The analysis showed that the needs for cooling will continue to increase and is expected that the situation will be deteriorated especially for low-income population.

Moreover passive convective cooling technologies have been tested in demonstration and real scale applications with excellent results. The efficiency of the proposed passive cooling systems is found to be high, while their environmental quality is excellent. Based on the research developments, many of the proposed systems and in particular the heat dissipation systems have been commercialised and are available to the public.

It is evident that further research is necessary in order to optimise the existing systems and develop new ones in order to relief the energy burden that is continuously increasing.

References

Akbari H, Davis S, Dorsano S, Huang J, Winnett S (1992) Cooling our communities: a guidebook on tree planting and light-colored surfacing. U.S. Environmental Protection Agency, Office of Policy Analysis, Climate Change Division, Washington, DC

Allard F, Santamouris M (eds) (1998) Natural ventilation in buildings: a design handbook. James & James (Science Publishers) Ltd, London

Antinucci M, Asiain D, Fleury B, Lopez J, Maldonado E, Santamouris M, . . . Yannas S (1992) Passive and hybrid cooling of buildings – state of the art. Int J Sol Energy 11(3–4):251–271

Asimakopoulos DA, Santamouris M, Farrou I, Laskari M, Saliari M, Zanis G, . . . Giannakopoulos C (2012) Modelling the energy demand projection of the building sector in Greece in the 21st century. Energy Build 49. doi:10.1016/j.enbuild.2012.02.043

Baccini M, Biggeri A, Accetta G, Kosatsky T, Katsouyanni K, Analitis A, . . . Michelozzi P (2008) Heat effects on mortality in 15 European cities. Epidemiology (Cambridge, MA) 19 (5):711–719. doi:10.1097/EDE.0b013e318176bfcd

Badescu V, Isvoranu D (2011) Pneumatic and thermal design procedure and analysis of earth-to-air heat exchangers of registry type. Applied Energy 88:1266–1280. doi:10.1016/j.apenergy.2010.10.019

Ballestini G, De Carli M, Masiero N, Tombola G (2005) Possibilities and limitations of natural ventilation in restored industrial archaeology buildings with a double-skin façade in Mediterranean climates. Build Environ 40(7):983–995

Blondeau P, Spérandio M, Allard F (1997) Night ventilation for building cooling in summer. Sol Energy 61(5):327–335

Breesch H, Bossaer A, Janssens A (2005) Passive cooling in a low-energy office building. Sol Energy 79:682–696

Cartalis C, Synodinou A, Proedrou M, Tsangrassoulis A, Santamouris M (2001) Modifications in energy demand in urban areas as a result of climate changes: an assessment for the southeast Mediterranean region. Energy Convers Manag 42(14):1647–1656. doi:10.1016/S0196-8904(00)00156-4

Duffy MJ, Hiller M, Bradley DE, Keilholz W, Thornton JW (2009) Trnsys – features and functionality. Energy 1950–1954. In: Eleventh international IBPSA conference, Glasgow, Scotland, 27–30 July 2009

Fiksel A, Thornton JW, Klein SA, Beckman WA (1995) Developments to the TRNSYS simulation program. J Sol Energy Eng 117(2):123–127. doi:10.1115/1.2870836

Finn DP, Connolly D, Kenny P (2007) Sensitivity analysis of a maritime located night ventilated library building. Sol Energy 81(6):697–710

Fintikakis N, Gaitani N, Santamouris M, Assimakopoulos M, Assimakopoulos DN, Fintikaki M, . . . Doumas P (2011) Bioclimatic design of open public spaces in the historic centre of Tirana, Albania. Sustain Cities Soc 1(1):54–62

Flourentzou F, Van Der Maas J, Roulet C-A (1998) Natural ventilation for passive cooling: measurement of discharge coefficients. Energy Build 27(3):283–292

Fraisse G, Boichot R, Kouyoumji J-L, Souyri B (2010) Night cooling with a ventilated internal double wall. Energy Build 42(3):393–400

Geros V, Santamouris M, Tsangrasoulis A, Guarracino G (1999) Experimental evaluation of night ventilation phenomena. Energy Build 29(2):141–154

Geros V, Santamouris M, Karatasou S, Tsangrassoulis A, Papanikolaou N (2005) On the cooling potential of night ventilation techniques in the urban environment. Energy Build 37 (3):243–257

Givoni B (1994) Passive and low energy cooling of buildings. Van Norstrand Reinhold, New York

Golneshan AA, Yaghoubi MA (1990) Simulation of ventilation strategies of a residential building in hot arid regions of Iran. Energy Build 14(3):201–205

Grosso M, Raimondo L (2008) Horizontal air to-earth heat exchangers in Northern Italy - Testing, design and monitoring. International Journal of Ventilation 7:1–10

Hassid S, Santamouris M, Papanikolaou N, Linardi A, Klitsikas N, Georgakis C, Assimakopoulos DN (2000) Effect of the Athens heat island on air conditioning load. Energy Build 32 (2):131–141. doi:10.1016/S0378-7788(99)00045-6

Jiru TE, Bitsuamlak GT (2010) Advances in applications of CFD to natural ventilation. In: The fifth international symposium on computational wind engineering, Chapel Hill, NC, USA, 23–27 May 2010

Kapsomenakis J, Kolokotsa D, Nikolaou T, Santamouris M, Zerefos SC (2013) Forty years increase of the air ambient temperature in Greece: the impact on buildings. Energy Convers Manag 74:353–365

Kolokotroni M (2001) Night ventilation cooling of office buildings: parametric analyses of conceptual energy impacts. ASHRAE Trans 107(1):479–489

Kolokotroni M, Aronis A (1999) Cooling-energy reduction in air-conditioned offices by using night ventilation. Appl Energy 63(4):241–253

Kolokotroni M, Tindale A, Irving S (1997) NiteCool: office night ventilation pre-design tool. In: 18th AIVC conference – ventilation and cooling, Athens

Kolokotroni M, Webb BC, Hayes SD (1998) Summer cooling with night ventilation for office buildings in moderate climates. Energy Build 27(3):231–237

Kolokotroni M, Giannitsaris I, Watkins R (2006) The effect of the London urban heat island on building summer cooling demand and night ventilation strategies. Sol Energy 80(4):383–392

Kovats RS, Hajat S, Wilkinson P (2004) Contrasting patterns of mortality and hospital admissions during heatwaves in London, UK. Occup Environ Med 61(11):893–898

Krüger E, González Cruz E, Givoni B (2010) Effectiveness of indirect evaporative cooling and thermal mass in a hot arid climate. Build Environ 45(6):1422–1433

Kubota T, Chyee DTH, Ahmad S (2009) The effects of night ventilation technique on indoor thermal environment for residential buildings in hot-humid climate of Malaysia. Energy Build 41(8):829–839

Li Y, Heiselberg P (2002) Analysis methods for natural and hybrid ventilation – a critical literature review and recent developments. Int J Vent 1(4):3–20, (Hybrid Ventilation Special Edition)

Li Y, Nielsen PV (2011) CFD and ventilation research. Indoor Air 21(6):442–453. doi:10.1111/j.1600-0668.2011.00723.x

Livada I, Santamouris M, Niachou K, Papanikolaou N, Mihalakakou G (2002) Determination of places in the great Athens area where the heat island effect is observed. Theor Appl Climatol 71 (3–4):219–230. doi:10.1007/s007040200006

Livada I, Santamouris M, Assimakopoulos M (2007) On the variability of summer air temperature during the last 28 years in Athens. J Geophys Res D Atmos 112, D12103. doi:10.1029/2006JD008140

Macias M, Mateo A, Schuler M, Mitre EM (2006) Application of night cooling concept to social housing design in dry hot climate. Energy Build 38(9):1104–1110

Manz H, Frank T (2005) Thermal simulation of buildings with double-skin façades. Energy Build 37(11):1114–1121

Maragogiannis K, Kolokotsa D, Maria EA (2011) Study of night ventilation efficiency in urban environment: technical and legal aspects. In: The 52nd international scientific conference of Riga Technical University Energy Resources 2011, Riga

McMichael AJ, Haines JA, Slooff R, Sari Kovats R (1996) Climate change and human health : an assessment/prepared by a task group on behalf of the World Health Organization, the World Meteorological Association and the United Nations Environment Programme. WHO, Geneva

Pantavou K, Theoharatos G, Mavrakis A, Santamouris M (2011) Evaluating thermal comfort conditions and health responses during an extremely hot summer in Athens. Build Environ 46 (2):339–344. doi:10.1016/j.buildenv.2010.07.026

Pfafferott J, Herkel S, Jäschke M (2003) Design of passive cooling by night ventilation: evaluation of a parametric model and building simulation with measurements. Energy Build 35 (11):1129–1143

Sakka A, Santamouris M, Livada I, Nicol F, Wilson M (2012) On the thermal performance of low income housing during heat waves. Energy Build 49:69–77

Santamouris M (2007) Heat island research in Europe – the state of the art. Adv Build Energy Res 1:123–150

Santamouris M, Assimakopoulos D (eds) (1997) Passive cooling of buildings. James & James, Science Publishers Ltd, London

Santamouris M, Kolokotsa D (2013) Passive cooling dissipation techniques for buildings and other structures: the state of the art. Energy Build 57:74–94. doi:10.1016/j.enbuild.2012.11.002

Santamouris M, Mihalakakou G, Argiriou A, Asimakopoulos D (1996) On the efficiency of night ventilation techniques for thermostatically controlled buildings. Sol Energy 56(6):479–483

Santamouris M, Papanikolaou N, Livada I, Koronakis I, Georgakis C, Argiriou A, Assimakopoulos DN (2001) On the impact of urban climate on the energy consumption of building. Sol Energy 70(3):201–216. doi:10.1016/S0038-092X(00)00095-5

Santamouris M, Kapsis K, Korres D, Livada I, Pavlou C, Assimakopoulos M (2007a) On the relation between the energy and social characteristics of the residential sector. Energy Build 39 (8):893–905. doi:10.1016/j.enbuild.2006.11.001

Santamouris M, Paraponiaris K, Mihalakakou G (2007b) Estimating the ecological footprint of the heat island effect over Athens, Greece. Clim Change 80(3–4):265–276. doi:10.1007/s10584-006-9128-0

Santamouris M, Pavlou K, Synnefa A, Niachou K, Kolokotsa D (2007c) Recent progress on passive cooling techniques. Advanced technological developments to improve survivability levels in low-income households. Energy Build 39(7):859–866

Santamouris M, Sfakianaki A, Pavlou K (2010) On the efficiency of night ventilation techniques applied to residential buildings. Energy Build 42(8):1309–1313. doi:10.1016/j.enbuild.2010.02.024

Springer DA, Rainer LI, Dakin WL (2005) Development and testing of an integrated residential night ventilation cooling system. ASHRAE Trans 111(Part 2):501–510

Stathopoulou E, Mihalakakou G, Santamouris M, Bagiorgas HS (2008) On the impact of temperature on tropospheric ozone concentration levels in urban environments. J Earth Syst Sci 117 (3):227–236. doi:10.1007/s12040-008-0027-9

Thiers S, Peuportier B (2008) Thermal and environmental assessment of a passive building equipped with an earth-to-air heat exchanger in France. Sol Energy 82:820–831. doi:10.1016/j.solener.2008.02.014

Trombe A, Pettit M, Bourret B (1991) Air cooling by earth tube heat exchanger: Experimental approach. Renewable Energy 1:699–707

Van der Maas J, Roulet C-A (1991) Nighttime ventilation by stack effect. ASHRAE Trans 7:516–524

Zimmerman M, Andersson J (1998) Low energy cooling – case study buildings. International energy agency, energy conservation in buildings and community systems programme. Annex 28 final report

Chapter 5
Thermal Comfort Inside and Outside Buildings

Richard de Dear and Jungsoo Kim

Abstract Engineers and architects are required to consider human occupants or pedestrians during the design process. The topic of human thermal comfort is often considered to be a long way from the traditional, hard science disciplines normally associated with engineering. Nevertheless there is a scientific basis for thermal comfort. The topic draws on several scientific disciplines, including physics (especially heat transfer and meteorology) and physiology. But most importantly of all, thermal comfort falls within the scope of psychology, since it is defined as 'that condition of mind that expresses satisfaction with the thermal environment'. Some general principles are relevant to the topic of thermal comfort both indoors and outdoors, but there are some very striking differences between the two settings as well, and these differences have significant implications for engineering of indoor and semi-outdoor climates.

Keywords Thermal comfort • PMV/PPD • HVAC • Adaptive comfort • Natural ventilation

5.1 Introduction: Thermal Comfort as a Design Consideration

Thermal comfort indoors has been the subject of scientific research ever since the ability to control indoor climate with air conditioning became commercially viable in the early decades of the twentieth century. There are six physical parameters that go into the comfort equation, comprising four environmental parameters (air temperature t_a, mean radiant temperature t_r, relative humidity rh and airspeed v) that affect heat transfers to or from the body and two personal parameters that describe the amount of heat being generated in the body and the resistance to that heat being lost to the environment (*met*, *clo*). From the designer's point of view, the four environmental parameters are affected by different elements of the built

R. de Dear (✉) • J. Kim
Faculty of Architecture, Design and Planning, University of Sydney, Sydney, NSW 2006, Australia
e-mail: richard.dedear@sydney.edu.au

Y. Tamura, R. Yoshie (eds.), *Advanced Environmental Wind Engineering*, DOI 10.1007/978-4-431-55912-2_5

environment and are therefore amenable to deliberate control by both architectural elements and mechanical services. The easiest to control is air temperature, and the mechanism of direct control is heating, ventilation and air conditioning (HVAC). Relative humidity is also subjected to control by HVAC systems, sometimes, but because of the enormous amount of energy required to chill supply air below its dew point just to remove moisture, this capability of HVAC is not commonly exploited. Mean radiant temperature (t_r) refers to the average surface temperature of the enclosure, as 'seen' by the occupants, and it controls the thermal (infrared) radiation exchange with the human body. This includes wall, floor, ceiling and window temperatures, which are not commonly controlled parameters in building. However, with sufficient insulation it is possible to maintain wall, floor and ceiling temperatures close to the air temperature prevailing within the space. Glazing, however, is a bit more problematic, and the modern tendency for architects to use this material across large expanses of perimeter walls (facades) poses complex thermal comfort problems for the occupants. If single-pane glazing is used, there is a high risk that the internal surface of the window will have a low surface temperature whenever outdoor air temperatures are low, giving rise to undesirable local cooling of the occupants (termed 'radiant draught'). When glazing is directly irradiated by the sun, there is a risk of another local discomfort – high-temperature radiant heat from the direct solar radiation penetrating the glazing and also from the glazing itself after it has been heated by absorbed solar radiation. Airspeed is the fourth environmental control of the human body's heat balance and is a parameter that can be manipulated to varying degrees by intelligent design of the size, placement and configuration of openings in the walls or ceiling. By consideration of the outdoor meteorological and climatic conditions surrounding the building (site climate), air movement within the occupied zone can be exploited as a low-energy strategy for cooling building fabric and also the building's occupants in hot climates. For mechanically regulated indoor climates, airspeed is partially controllable by placement and sizing of air inlets relative to the occupants of the space and also the velocity of supply air as it enters the occupied zone. However, for most HVAC design systems in use today, airspeed within the occupied zone is not readily controlled, and so it is generally constrained to barely detectible speeds ($<$2 m/s) so as to avoid the risk of draught in cooler-than-neutral indoor climates. However, newer approaches to HVAC such as displacement ventilation rely on the delivery of very low-speed conditioned air to the space just above floor level where it accumulates and rises slowly due to thermal buoyancy as it becomes warmed by various heat sources within the occupied zone (occupants, computers, lighting, etc.).

The other two factors controlling the human body's heat balance are referred to as the personal comfort parameters – metabolic rate (*met*) and clothing insulation (*clo*). Since these two relate to the behaviour of the occupants of a building rather than the building itself, they are generally regarded as beyond the designer's control. By thoroughly understanding the nature of the occupancy of the building in question – learning about the dress codes or customs of the building's occupants and the types of activity levels they are likely to engage in within the building – the designer should be able to make reasonable estimates of *clo* and *met* values. This knowledge can then be used to optimise the design of the other four environmental comfort parameters so that occupant thermal comfort can be achieved.

Designing for comfort in *outdoor* settings is not as straightforward as the indoor scenario described above for the simple reason that these settings are not fully enclosed, and therefore, total control of the four environmental comfort parameters is not feasible. Whilst control of space temperature may not be a design option, the remaining three parameters can often be attenuated if not totally controlled. Air-speeds can be reduced by wind shelters or increased with deliberate design of 'wind catchers' or mechanical fans. Dehumidification may not be feasible in hot and humid outdoor settings, but humidification with sprays of droplets is one option in arid climate zones, with the added advantage of localised evaporative cooling. Radiant temperatures outdoors cannot be directly controlled, but solar shading is an essential strategy for keeping radiant loads within an acceptable range for the occupants of hot outdoor and semi-outdoor settings.

5.2 Thermal Comfort Indoors

ASHRAE Standard 55-2013 (ASHRAE 2013) currently recommends two distinct strategies for defining the indoor comfort zone, depending on whether the setting is (a) a mechanically air-conditioned building where occupants typically have little or no access to thermal controls or (b) a naturally ventilated building where occupants have access to operable windows. In relation to the former setting – centrally controlled air conditioning, the 'father of modern thermal comfort science', Professor P.O. Fanger (1970), subscribed to the adage that you 'can't please all of the people all of the time', so he devised an index (*PMV*/*PPD*) to maximise comfort for multiple occupants sharing the same indoor climatic conditions. Predicted Mean Vote (PMV) is an index that predicts the mean thermal sensation of a group of people, determined from thermal the six comfort parameters introduced earlier (i.e. t_a, t_r, *rh*, *v*, *met* and *clo*). The result is output on a seven-point scale ranging from −3 to +3 (−3 *cold*, −2 *cool*, −1 *slightly cool*, 0 *neutral*, +1 *slightly warm*, +2 *warm*, +3 *hot*). *PPD* (Predicted Percentage Dissatisfied) predicts the percentage of dissatisfied with thermal conditions and is derived directly from the *PMV* index. In ASHRAE Standard 55 (ASHRAE 2013), the acceptable thermal condition, so-called comfort zone , is defined as *PMV* values between ±0.5, corresponding to a *PPD* value of less than 10 %. By making an assumption that there will typically be an additional 10 % dissatisfied from local discomforts such as draught, radiant asymmetry and vertical air temperature stratification, the practising engineer can assume that an internal space with a PMV of ±0.5 will achieve a combined level of dissatisfaction below 20 %, commonly regarded as an industry benchmark. *PPD* reaches its minimum value (5 %) when *PMV* equals zero (i.e. *neutral*), then *PPD* starts to increase as *PMV* values deviates from neutrality in either the warm or cool direction.

The underlying concept of the *PMV*/*PPD* indices is Fanger's first prerequisite for thermal comfort – net heat balance for the human body. The simple heat equation identifies the state of thermal comfort as occurring when heat fluxes into the body by metabolism, convection, radiation and conduction are balanced by the fluxes out of the body via work, convection, radiation, evaporation and conduction. Fanger's second requirement for comfort was that mean skin temperature and evaporative heat loss from the skin are within their respective 'comfort ranges'. Comfort ranges for skin temperature and evaporative heat loss were defined empirically as functions of metabolic rate. The comfortable skin temperature is an inverse function of metabolic rate and the comfortable evaporative heat loss is a positive function of metabolic rate.

Referring back to the ASHRAE 55-2013 (ASHRAE 2013) comfort standard again, the second recommended approach is that for naturally ventilated spaces where there is no mechanical cooling available, but occupants have direct access to operable windows. In this setting, the ASHRAE standard prescribes de Dear and Bragers' (1998) adaptive approach to comfort, which differs fundamentally from Fanger's heat-balance approach in that it accepts non-heat-balance factors such as culture, thermal history, context and thermal expectation as exerting influence over thermal comfort outcomes. The adaptive model is based on extensive field studies of thermal comfort drawn from a wide spectrum of climate zones and 160 different buildings across the globe (de Dear 1998) and indicates that thermal comfort can be attained in warmer indoor temperatures if the outdoor climate is warm and cooler indoor temperatures when the outdoor climate is cold. In effect, the occupants of naturally ventilated buildings adapt to the conditions they are exposed to by a variety of mechanisms, including physiological (acclimatisation), behavioural (using operable windows) and psychological (adjusting comfort expectations). The major conceptual departure of the adaptive model is its reference to thermal history, expectations and attitudes, perceived control and availability of behavioural thermoregulatory options (sometimes referred to as 'adaptive opportunities'). Whilst the heat-balance model is able to account for some degree of behavioural adaptation, such as changing one's clothing or adjusting local air velocity, it ignores the psychological dimension of adaptation completely, which explains the inconstant predictive skill of PMV/PPD. Psychological dimensions of thermal adaptation may be particularly important in contexts where people's interactions with the environment (i.e. personal thermal control through adaptive opportunities), or diverse thermal experiences, may alter their expectations and, thus, their thermal sensation and satisfaction. Figure 5.1 shows the upper and lower operative temperature limits for 80 and 90 % acceptability, defined by adaptive model. The running 7-day mean outdoor air temperature ($T_{a(out)}$) is a basis used to define the adaptive driver for input to the adaptive comfort model, which is expressed as

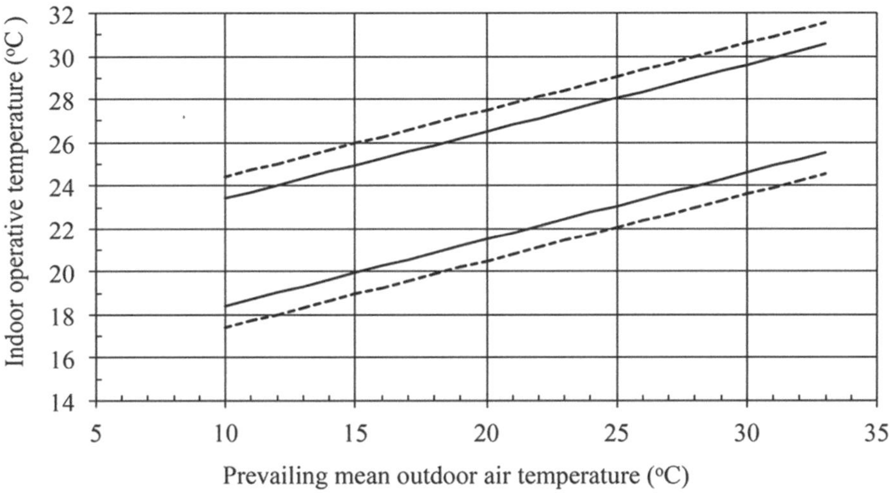

Fig. 5.1 Acceptable operative temperature (t_o) ranges for naturally conditioned spaces, *solid lines* and *dashed lines* represent 90 % and 80 % acceptability limits, respectively (ASHRAE 2013; de Dear and Brager 1998)

$$T_{a(out)} = 0.34T_{a(day\text{-}1)} + 0.23T_{a(day\text{-}2)} + 0.16T_{a(day\text{-}3)} + 0.11T_{a(day\text{-}4)} + 0.08T_{a(day\text{-}5)} + 0.05T_{a(day\text{-}6)} + 0.03T_{a(day\text{-}7)}$$

The 7-day running mean decay function signifies that the mean daily outdoor temperature yesterday is the most important driver of adaptive comfort (34 % weighting), followed by the day before yesterday (23 % weighting), then the day before that (16 % weighting) and so on back to 7 days ago which carries just 3 % weighting. These exponentially decaying weighting coefficients are based on empirical evidence provided by a clothing behaviour observational study (Morgan and de Dear 2003) in a shopping mall in Sydney, Australia. In that study, a detailed *clo* estimate was made for a random sample of about 50 subjects each day, continuously for a 6-month period. The daily mean *clo* value was then correlated with the mean outdoor temperature on the preceding day (day $x-1$). That correlation was repeated for *clo* on mean outdoor temperature on days $x-2, x-3, x-4, x-5, x-6$ and $x-7$. The relative sizes of those correlation coefficients formed the basis of the exponentially decaying weighting coefficients appearing in the above equation.

As implied by the adaptive model, psychosocial factors such as occupants' thermal history, expectations and adaptive behaviours can all play an important role in shaping their comfort zone inside buildings. Kim and de Dear (2012) explored differences between occupants in air-conditioned buildings and naturally ventilated buildings, focusing on their expectations of what indoor thermal environment should be like. The analysis was based on a large post-occupancy evaluation database ($n = 22{,}518$ from 137 office buildings) (Zagreus et al. 2004), and it suggested that people in air-conditioned buildings tended to have higher expectations for thermal uniformity inside their buildings, becoming highly critical whenever a building failed to meet those expectations (i.e. when indoor temperature deviates from the narrow temperature range they have come to expect). In contrast, occupants in naturally ventilated buildings reported relatively modest expectations of indoor thermal environment, tending to accept less-than-ideal thermal conditions. Therefore the researchers contended that thermal comfort in naturally ventilated buildings was regarded by their occupants as a 'bonus'. Such a building is 'forgiven' by occupants whenever it is thermally underperforming, but it can also be pleasantly surprising whenever it exceeds occupants' 'thermal expectations'. The air-conditioned subsample gave a different response – thermal comfort was perceived as 'basic' or minimum requirement of occupants of air-conditioned buildings.

In a centrally air-conditioned building, creating thermally uniform and imperceptible indoor environments (neither warm nor cool, i.e. neutral) based on *PMV/PPD* model has been the design objective for building services engineer since the invention of air conditioning. On the other hand, occupants in naturally ventilated buildings are exposed through operable windows to various unquantifiable or qualitative sensations (e.g. nice breeze, connection to outdoor weather rhythms, etc.), which can never be experienced in a sealed facade air-conditioned space. Whilst 'not being uncomfortable' is as good as it gets in air-conditioned buildings, a more dynamic and variable indoor climate, sometimes departing from neutrality in naturally ventilated buildings, may let occupants perceive thermal comfort transcending basic functional needs (Kim and de Dear 2012). According to the alliesthesia hypothesis, a dynamic indoor thermal climate offers the prospect of thermal pleasure, perhaps even thermal delight (Heschong 1979).

Since heating, ventilation and air conditioning (HVAC) typically account for about half of a commercial building's energy usage, the naturally ventilated approach is attracting renewed interest from architects and engineers who are keen to reduce the greenhouse gas footprint of the buildings they design. These designs can be fully naturally ventilated, or mixed-mode, depending on the severity of the external climatic environment. A mixed-mode building is one that is naturally ventilated for some of the time (weather permitting) using operable windows or vents whenever the external climate is favourable and air-conditioned whenever external atmospheric conditions are unable to deliver comfortable internal conditions. By intermittently flipping into natural ventilation mode, mixed-mode (or hybrid) ventilation buildings can dramatically reduce their energy use associated with HVAC systems whilst still maintaining the quality and acceptability of their indoor environment. Besides the impact of operating mode on energy consumption, there are differences in occupants' perception of thermal environment as well. According to a post-occupancy survey conducted in a mixed-mode building with accompanying physical measurements, the same occupants in the same building responded differently to the identical thermal environment, depending on the building's ventilation mode (Deuble and de Dear 2012). Within naturally ventilated mode, the occupants were more accepting of thermal conditions that were 'warmer than neutral', compared to when the building was operating under air-conditioned mode.

5.3 Thermal Comfort Outdoors

Thermal comfort research outdoors and in 'halfway' settings, sometimes called 'semi-outdoors', is usually based on the same four microclimatic parameters (t_a, t_r, rh and v) as indoor comfort research. However one of the key parameters, mean radiant temperature, is fundamentally different in the outdoor and semi-outdoor setting because there is a complex array of both short- and long-wave radiation fluxes impinging on the subject from all directions as a result of sunshine (which is generally not relevant indoors). A software tool by Matzarakis et al. (2010) estimates outdoor mean radiant temperature for practical applications.

Once the four microclimatic parameters (t_a, t_r, rh and v) have been directly measured or estimated in the outdoor or semi-outdoor setting, they are typically integrated together with estimates of clothing insulation (*clo*) and metabolic rates (*met*) into a heat-balance index of thermal comfort such as *OUT_SET*, *PT*, *PET* or, most recently, the Universal Thermal Climate Index (*UTCI*) (Hoeppe 2002). These indices integrate the effects of all six comfort parameters on the human thermal balance – giving a single number (environmental rating) that purport to represent a metric of subjective thermal comfort experienced by the 'average' person exposed to those outdoor environmental conditions.

Case studies of research in sports stadia were described by Bouyer et al. (2007). Wind environmental data in this study were estimated from wind tunnel simulations of scale model, whilst other comfort model inputs were estimated from meteorological records or dynamic thermal simulations of the structure's response to typical meteorological conditions.

A significant theoretical critique of this approach focuses on the presumption that 'neutral' or 'comfortable' means the same thing in physical or physiological terms in an *outdoor* context as it does indoors. There is no empirical basis for this extrapolation from indoor research to outdoors. As addressed in the earlier section, only a narrowband close to neutrality (i.e. $-0.5 < PMV < +0.5$) within boundaries of −3 (*cold*) and +3 (*hot*) is deemed as thermally acceptable range for air-conditioned spaces. However, outdoor climate conditions are expected to deviate more significantly from neutrality, covering a much wider range than typical indoor climates because the latter are commonly controlled or moderated by architectural or mechanical systems. Therefore the thermally acceptable range in outdoor spaces as well can be expected to stretch further from $PMV = 0$. Indeed, there are detailed outdoor and semi-outdoor comfort studies in Sydney (Spagnolo and de Dear 2003) and in Tokyo (Nakano and Tanabe 2004) indicating a discrepancy of up to 4 °C or 5 °C between optimum indoor and outdoor comfort conditions.

Using the classic thermal comfort field research design (the 'right-here-right-now' survey design), Spagnolo and de Dear (2003) conducted an outdoor comfort survey in various settings in subtropical Sydney with a purpose of developing thermal comfort instruments and protocols for (semi)-outdoor research. A mobile meteorological station was developed to measure outdoor microclimates including t_a, *rh*, *v*, *global short-wave radiation (k)*, *diffuse short-wave radiation (d)* and *long-wave radiation (l)*. A distinctive feature of this instrumentation was its alternative approach to measurement of mean radiant temperature (t_r). The sensors were installed on a hinged plate that could be inverted to record radiative fluxes from both hemispheres, one facing upwards and the other downwards (Fig. 5.2 left). Use of radiation sensors that integrate across 180° angle and two hemispheres provided a reliable average of the short- and long-wave radiation fluxes from all directions, which in turn simplified t_r calculation greatly. Their 'right-here-right-now' comfort survey carried out in diverse outdoor and semi-outdoor spaces with different surface types and human biometeorological conditions indicated that the comfort zone prescribed for indoor spaces is not directly applicable to outdoor contexts. The study reported a much larger range of comfort zone in outdoor settings than indoors, possibly because of the much great thermal variability of outdoor environment over space and time being factored into people's expectations.

A more recent study conducted in Japan took a completely different approach to the task of measuring human biometeorological urban climates. Instead of measuring at a single point in time and space, their focus was on the traverse of pedestrians through time and space. This required a new method of sensing the main comfort parameters and also the corresponding human physiological responses to those parameters. As illustrated in Fig. 5.2 (right panel), Nakayoshi et al. (2015) developed compact, wearable instrumentation that measures both meteorological variables (t_a, *rh*, *v*, *short-wave radiation s* and *long-wave radiation l*) and simultaneous physiological variables (t_{skin}, *pulse rate* and *body motion*). Whilst the subjects were walking through a series of urban outdoor settings, the microclimate enveloping the subject and the matching physiological states were simultaneously scanned into a logger worn by each subject. Some other physiological indices such as tympanic temperature and sweat rate were manually measured intermittently by the

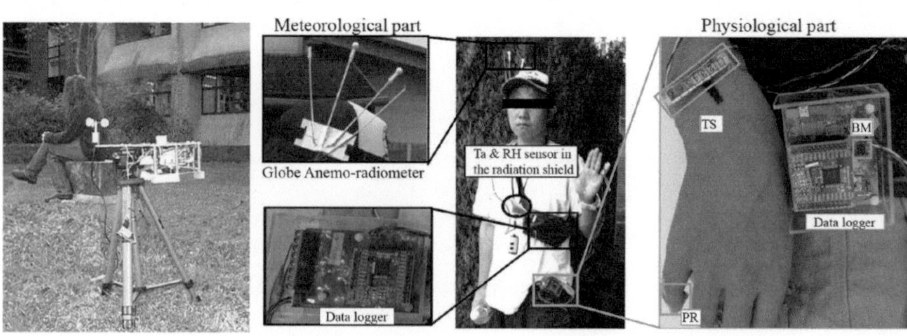

Fig. 5.2 Mobile meteorological station by Spagnolo and de Dear (*left*) (Spagnolo and de Dear 2003) and wearable thermal physiology measurement system by Nakayoshi et al. (*right*) (*TS* thermal sensation, *PR* pulse rate, *BM* body motion detected by accelerometers) (Nakayoshi et al. 2015)

researchers throughout the experiment (Nakayoshi et al. 2015). Given that thermal comfort studies in outdoor settings have previously been restricted by the difficulties of data collection, particularly if the environment is transient as is the case for pedestrians on the move, the instrumentation and protocol proposed by Nakayoshi et al. make a significant improvement in the practicality of characterising objective urban biometeorological environments and simultaneous subjective comfort responses of pedestrians moving through those environments.

Divergence between indoors and outdoor thermal comfort zones was highlighted by another *right-here-right-now* comfort survey carried out in humid tropical climate of Singapore (Yang et al. 2013). The naturally hot and humid climatic conditions of Singapore have been exacerbated in recent decades by an urban heat island resulting from rapid urbanisation. Thermal comfort perception and preference of people in various outdoor spaces (typically resting places) at various times of a day (morning, midday, afternoon and evening) were investigated matched against simultaneous measurements of the key microclimatic parameters (using an identical field method to that described earlier in Spagnolo and de Dear (2003)). The neutral operative temperature and preferred temperature based on 2036 sets of comfort questionnaires were estimated as 28.7 °C and 26.5 °C, respectively. In Singapore, the operative temperature range of 26–32 °C was found to be *acceptable* to people in outdoor spaces.

A consistent finding in all of these applications of the conventional *right-here-right-now* thermal comfort research method to outdoor settings is a discrepancy between the temperatures that indoor heat-balance models predict should be comfortable and what subjects in outdoor settings actually say is comfortable. This offset between indoor and outdoor comfort has usually been explained in terms of differences in comfort expectations (Spagnolo and de Dear 2003; Yang et al. 2013), but an alternative psychophysiological hypothesis has emerged in recent years (de Dear 2011) that gives a more thorough explanation of why positive feelings of comfort can be experienced in thermal situations that would not normally be regarded as comfortable were they to be encountered in indoor settings. Called 'alliesthesia', this hypothesis points, in a hot environmental context, to the effects of a cooling breeze on the skin and cutaneous cold thermoreceptors just beneath the skin surface. Even though the transient cooling may not be sufficient to fully restore skin temperature and skin latent heat loss to the values corresponding to 'neutral' for their metabolic rate (according to Fanger's indoor heat-balance comfort theory), the skin cooling is sufficient to trigger the impression that restoration of neutrality is on its way. Transient local cooling at skin surface in hot and humid thermal environments can be perceived as something more pleasant than a 'neutral' thermal sensation of the type encountered, for example, in a conventionally air-conditioned office environment. Alliesthesia also explains why a glass of water is being especially enjoyable when we are slightly dehydrated or why a meal tastes particularly delicious when we are hungry. The same meal taken when we are sated would taste ordinary by comparison – even though the meal itself is identical in both situations. The significance of alliesthesia in this chapter on thermal comfort is that indoor or outdoor thermal environments may not need to comply with the very narrow range prescribed by the *PMV*/*PPD* index in order to be acceptable. Indeed, warm indoor or outdoor environments can still be rated as acceptable by their occupants, as long as there is some relief provided by air movement around the body. That air movement can be mechanically generated by fans or it can result from the movement of a pedestrian *through* warm or even hot urban microclimates (Nakayoshi et al. 2015) – whenever localised cooling is detected on the skin surface, the alliesthesial response of thermal pleasure will be elicited in a pedestrian having a whole-body thermal sensation that is warmer than neutral.

5.4 Conclusions

Thermal comfort is that condition of mind that expresses satisfaction with the thermal environment. It can be predicted in air-conditioned environments by the *PMV*/*PPD* index using just four environmental measurements and two personal parameters (*clo* and *met*). The goal of design in such contexts is to maintain *PMV* equal to or very close to zero (*neutral*) for as many of the occupied hours of the building's life as possible.

The range of thermal conditions required for comfort in naturally ventilated buildings is considerably wider than in air-conditioned buildings. The adaptive comfort theory describes an indoor comfort zone that drifts upwards in buildings located in hot climates and downwards in cold climate zones. This observation carries significant implications for the design and operation of low carbon-emitting buildings, and this represents one of our more promising avenues for mitigating humankind's impact on global climate.

Comfort in the outdoor and semi-outdoor climatic contexts is controlled by the same factors as described above for indoor settings, but cultural and contextual factors, including expectations, appear to widen the acceptable comfort zone even further. Rather than striving for static, neutral thermal conditions in outdoor settings, the designer should consider natural microclimatic dynamics as a positive attribute. In warm to hot microclimates, the key to creating comfort for the occupants seems to be the exploitation of wind (breeze) resources.

References

ASHRAE (2013) ANSI/ASHRAE standard 55-2013: thermal environmental conditions for human occupancy. American Society of Heating, Refrigerating and Air-Conditioning Engineers, Inc., Atlanta

Bouyer J, Vinet J, Delpeche P, Carre S (2007) Thermal comfort assessment in semi-outdoor environments: application to comfort study in stadia. J Wind Eng Ind Aerodyn 95:936–976

de Dear R (1998) A global database of thermal comfort field experiments. ASHRAE Trans 104 (1b):1141–1152

de Dear R (2011) Revisiting an old hypothesis of human thermal perception: alliesthesia. Build Res Inf 39(2):108–117
de Dear R, Brager G (1998) Developing an adaptive model of thermal comfort and preference. ASHRAE Trans 104(1):145–167
Deuble MP, de Dear R (2012) Mixed-mode buildings: a double standard in occupants' comfort expectations. Build Environ 54:53–60
Fanger PO (1970) Thermal comfort. Danish Technical Press, Copenhagen
Heschong L (1979) Thermal delight in architecture. MIT Press, Cambridge, MA
Hoeppe P (2002) Different aspects of assessing indoor and outdoor thermal comfort. Energy Build 34:661–665
Kim J, de Dear R (2012) Impact of different building ventilation modes on occupant expectations of the main IEQ factors. Build Environ 57:184–193
Matzarakis A, Rutz F, Mayer H (2010) Modelling radiation fluxes in simple and complex environments: basics of the RayMan model. Int J Biometeorol 54(2):131–139
Morgan C, de Dear R (2003) Weather, clothing and the thermal adaptation to indoor climate. Clim Res 24(3):267–284
Nakano J, Tanabe S (2004) Thermal comfort and adaptation in semi-outdoor environments. ASHRAE Trans 110:543–553
Nakayoshi M, Kanda M, Shi R, de Dear R (2015) Outdoor thermal physiology along human pathways: a study using a wearable measurement system. Int J Biometeorol 59(5):503–515
Spagnolo J, de Dear R (2003) A field study of thermal comfort in outdoor and semi-outdoor environments in subtropical Sydney Australia. Build Environ 38(5):721–738
Yang W, Wong NH, Jusuf SK (2013) Thermal comfort in outdoor urban spaces in Singapore. Build Environ 59:426–435
Zagreus L, Huizenga C, Arens E, Lehrer D (2004) Listening to the occupants: a web-based indoor environmental quality survey. Indoor Air 14(8):65–74

Chapter 6
Pedestrian Wind Environment Around Tall Buildings

Ted Stathopoulos and Bert Blocken

Abstract Pedestrian-level wind conditions around tall buildings are described by examining the aerodynamics of the urban environment and the various wind comfort criteria established in the wind engineering field. Experimental and, possibly, computational assessment of pedestrian-level wind conditions in the urban environment are described in detail. Particular emphasis has been placed on the state of the art and the capabilities of Computational Wind Engineering to determine at least mean values of wind speeds in the vicinity of buildings in urban areas. An approach toward the establishment of an overall comfort index taking into account, in addition to wind speed, the temperature, and relative humidity in the urban area under consideration, is presented.

Keywords Building • CFD • Comfort • Environment • Pedestrian • Wind tunnel

6.1 Introduction

The quality of open urban spaces has received a lot of attention in recent years. There is a broad recognition that microclimatic conditions contribute to the quality of life in cities, both from the economic as well as from the social viewpoint. Consequently, universities and other research organizations, municipal and other government forms, as well as construction and architectural companies have expressed significant interest and allocated resources to examine microclimatic conditions, particularly the effect of wind, on the outdoor human comfort. This task is quite complex because, contrary to the more or less controllable indoor comfort conditions, outside human comfort in an urban climate is generally

T. Stathopoulos (✉)
Department of Building, Civil and Environmental Engineering, Concordia University, Montreal, QC H3G 1M8, Canada
e-mail: statho@bcee.concordia.ca

B. Blocken
Department of the Built Environment, Eindhoven University of Technology, 5600 MB Eindhoven, The Netherlands

Department of Civil Engineering, Leuven University, 3001 Leuven, Belgium
e-mail: b.j.e.blocken@tue.nl; bert.blocken@bwk.kuleuven.be

Y. Tamura, R. Yoshie (eds.), *Advanced Environmental Wind Engineering*,
DOI 10.1007/978-4-431-55912-2_6

affected by a wide range of variables such as wind speed and direction, air temperature, solar radiation, possible precipitation in various forms, and the like.

In Europe, a 3-year (2001–2004) EU-funded project with extensive surveys carried out at different open spaces has carried out approximately 10,000 interviews (http://alpha.cres.gr/ruros). The project produced an urban design tool that provides architects, engineers, urban planners, and other decision-makers with means to assess effectively the construction of new buildings and the development of cities from the economic, psychophysiological, and sociological perspective of human comfort, air temperature, relative humidity, solar radiation, air quality, and human activity. Work has also been carried out within the auspices of the European Action C14 dealing with *Impact of Wind and Storm on City Life and Built Environment,* with a working group interested in the effects of wind on pedestrians, their assessment and comparisons, as well as the parameters that influence human comfort and its evaluation. Results have appeared in the 2002 Workshop in Nantes, e.g., Westbury et al. (2002) and in the International Conference in Urban Wind Engineering and Building Aerodynamics organized by the von Karman Institute for Fluid Dynamics in May 2004. In addition, the American Society of Civil Engineers (ASCE) has put out a pertinent state-of-the-art document (2003) which was developed with input from the European Action C14.

This chapter describes the aerodynamics of the urban environment and the reasons causing high wind speeds at sidewalks and, consequently, potential discomfort or danger to pedestrians, particularly in the proximity of tall buildings; it addresses the experimental and computational evaluations of the wind on people in the urban environment and focuses on the state of the art of the development of human outdoor comfort criteria by considering a wide range of parameters, including wind speed, air temperature, relative humidity, solar radiation, air quality, human activity, clothing level, age, and the like.

6.2 Aerodynamics of the Urban Environment

Strong winds are usually accelerated at the pedestrian level within the urban environment, say around tall buildings, due to particular aerodynamic configurations generally associated with tall buildings. In the case of a simple rectangular tall building, it is the boundary layer flow that causes descending flows toward the street level due to the pressure differences created by the velocity differences between higher and lower levels. This downflow is significant due to the pressure proportionality to the **square** of the velocity (Bernoulli equation) and the increase of the latter with the building height. Figure 6.1 demonstrates this effect, which is termed in the literature as *downwash*. Clearly, downwash is diminished drastically in the absence of boundary layer flow, and this explains the lack of adequate representation of wind effects in the building environment for simulations carried out in the past using aeronautical wind tunnels for building aerodynamics applications.

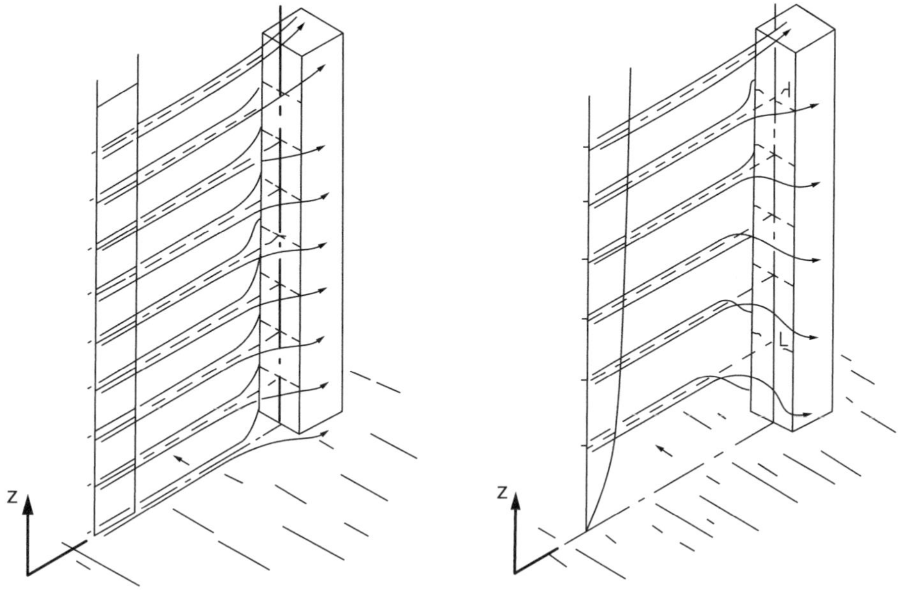

Fig. 6.1 Uniform and boundary layer wind flow around a tall rectangular building

In general, buildings will only induce high wind speeds at lower levels if a significant part of them is exposed to direct wind flows. It is actually the direct exposure to wind rather than building height alone which causes the problem. This is shown diagrammatically in Fig. 6.2. Another type of pedestrian-level winds is formed when high-speed winds pass through openings between high-pressure air on the windward wall and low-pressure regions on the leeward side of a building. Once more, the fair character of nature, which does not like pressure differences, prevails and strong flow is induced to correct the problem. Pedestrians in arcades of commercial buildings can testify regarding this situation, which is unpleasant to the store owners in these areas as well. Figure 6.3 shows this type of configuration along with other flow-induced mechanisms creating disturbances to the urban environment in the vicinity of buildings. These include but they are not limited to the effects of the large standing vortex in front of a building, the vortex flows generated after the flow separates and accelerates along the building front edges, and the wake-induced disturbances via the interaction of the flow coming from the building side faces and the recirculation flow regime created by the shear layer flow above the building. Clearly, wind direction is a significant factor here, in addition to the magnitude of the oncoming wind speed.

Additional common building configurations and potential influences on pedestrian-level winds are shown in Fig. 6.4 taken from Cochran (2004). These configurations include the effects of canopies, which may act as deterrents to the strong downflow prior to impacting on sidewalks or other pedestrian-free access

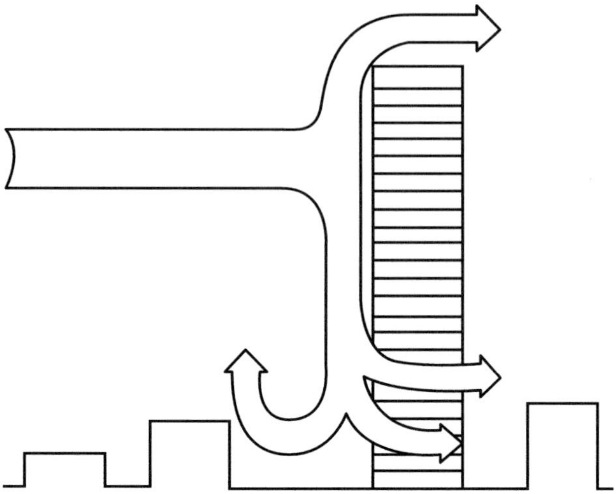

Fig. 6.2 Wind flow around buildings significantly taller than their surroundings (After Cochran (2004))

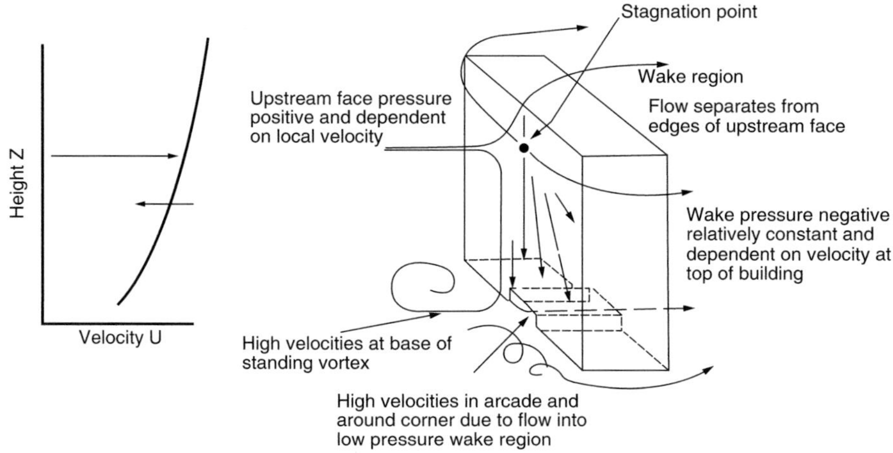

Fig. 6.3 Interaction of building with oncoming wind flow

areas around the building. However, such measures may create other problems by deflecting the wind from, say, a building entrance to another area around the building corners or across the street. Setbacks on the building surfaces or penthouses are elements generally remediating the pedestrian-level winds and are used rather extensively. Furthermore, a podium not intended for long-term pedestrian activities or vegetation in terms of bushes and coniferous-evergreen trees can also be used as a positive measure to amend harsh wind conditions at pedestrian level. Porous screens are also successful in deflecting winds without relocating the

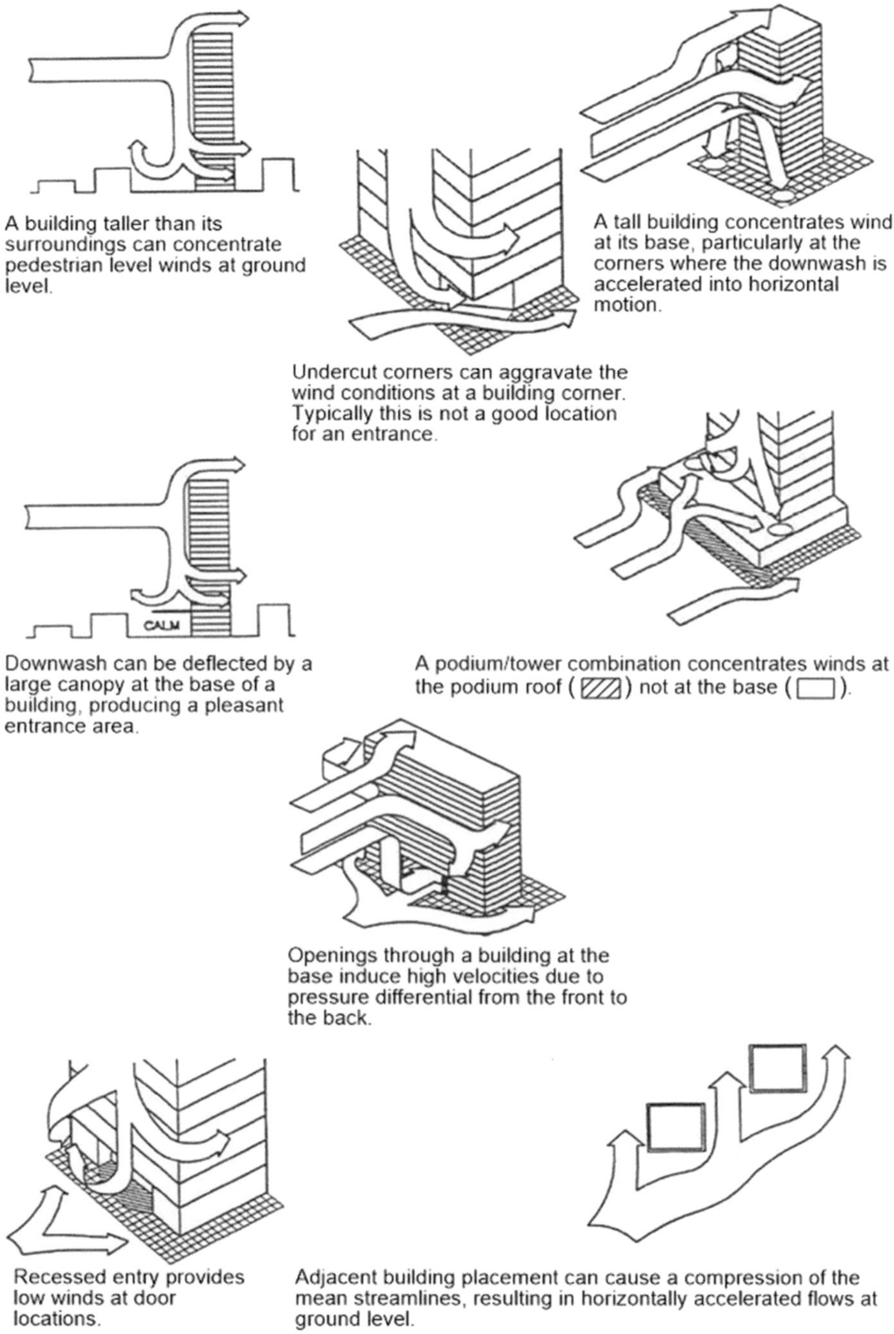

Fig. 6.4 Design features to change and/or ameliorate pedestrian wind conditions (After Cochran (2004))

adverse conditions on other places. An entrance alcove, as well as balconies, also diminishes sidewalk winds in the cities. However, high winds may be transferring on balconies themselves, particularly those near the edges of the building facades.

The previous discussion is really about isolated and mainly rectangular buildings. Curved buildings such as cylindrical shapes generally promote lateral flow, so they behave better as far as effects of pedestrian-level winds are concerned. Channeling effects appearing in the case of two or more buildings are generally critical, particularly if the wind direction is along the street or corridor formed between the buildings. Previous computational and experimental research by Blocken et al. (2007, 2008a, b) indicated that the channeling effects are particularly pronounced in the passage near ground-level, while at higher positions in the passage, the flow speed actually reduces compared to free-field conditions. As a result, the overall flow rate through the passage decreases compared to open-field conditions, but the flow speed at pedestrian level increases. Based on these findings, Blocken et al. (2007, 2008b) preferred the term "channeling effect" to be used instead of Venturi effect, as the Venturi effect strictly only applies to confined flows (Venturi 1799), while atmospheric flows are open flows.

If the wind conditions with one or two simple-shaped buildings in place can become so complex, one can easily imagine what would really happen with buildings of complex shapes interacting with the wind flow passing among them, particularly when the effect of ground topography and all adjacent buildings is taken into account. The problem becomes really difficult and for a number of years could only be solved experimentally via appropriate simulation in a boundary layer wind tunnel. However, with the significant progress in computational technology, attempts were made to address the problem of pedestrian-level winds in the urban environment computationally. More detailed discussion on the state of the art of computational fluid dynamics (CFD) applied to – among others – the study of pedestrian-level winds can be found in the review papers by Stathopoulos (1997; 2002) and in the review paper by Blocken (2014) on 50 years of Computational Wind Engineering.

Regardless of the approach used to determine the impact of wind flows at the pedestrian level, the previous comments have demonstrated that the direction of the oncoming wind together with its magnitude, i.e., speed, will be of paramount importance. If the wind climate in a city is distinctly directional, i.e., strong winds come always from a particular narrow fetch, it is clear that this set of directions should be really scrutinized because, in all likelihood, critical results will occur when the wind comes from these particular directions.

As an example, the basic wind environment of Montreal in terms of wind speeds and probabilities of exceedance from different directions is presented in Fig. 6.5. As clearly shown, westerly and southwesterly winds dominate while north and northeasterly winds may also be high. Note that these are upper level winds and significant changes may occur near the ground areas. In addition, differences exist between summer and winter wind data. Maximum summer winds are dominant from west, while winter winds are certainly higher and they blow primarily from southwest. In the great majority of pedestrian wind studies carried out for tall buildings in Montreal, it has been found that winds from the west/southwest and, to

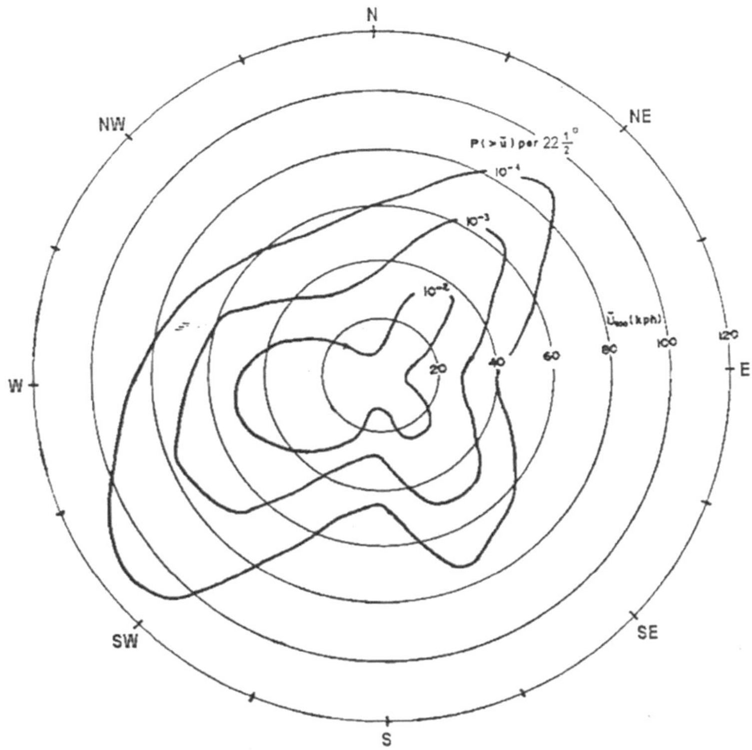

Fig. 6.5 Probability distributions of hourly mean wind speed at 300 m overground for daylight hours during the winter (derived from 10-year record of wind data obtained at a height of 10 m at Montreal's Trudeau Airport)

a lesser extent from the northeast, have produced the most critical adverse conditions.

In summary, there are two main flow types causing high pedestrian-level winds in the urban environment: downwash flows and horizontally accelerated flows. The former are diminished by podia and architectural features such as setbacks, balconies, and the like; the latter are ameliorated by alcoves, chamfered corners, landscaping (vegetation), or porous screens.

6.3 Wind Comfort Criteria

Several criteria have been developed in the wind engineering community for evaluating only the wind-induced mechanical forces on the human body and the resulting pedestrian comfort and safety. There are significant differences among the criteria used by various countries and institutions to establish threshold values for

Table 6.1 Extended land Beaufort scale showing wind effects on people (Lawson and Penwarden 1975; Isyumov and Davenport 1975)

Beaufort number	Description	Wind speed (m/s) at 1.75 m height	Effect
0	Calm	0.0–0.1	
1	Light air	0.2–1.0	No noticeable wind
2	Light breeze	1.1–2.3	Wind felt on face
3	Gentle breeze	2.4–3.8	Hair disturbed, clothing flaps, newspaper difficult to read
4	Moderate breeze	3.9–5.5	Raises dust and loose paper, hair disarranged
5	Fresh breeze	5.6–7.5	Force of wind felt on body, danger of stumbling when entering a windy zone
6	Strong breeze	7.6–9.7	Umbrellas used with difficulty, hair blown straight, difficult to walk steadily, sideways wind force about equal to forward walking force, wind noise on ears unpleasant
7	Near gale	9.8–12.0	Inconvenience felt when walking
8	Gale	12.1–14.5	Generally impedes progress, great difficulty with balance in gusts
9	Strong gale	14.6–17.1	People blown over by gusts

tolerable and unacceptable wind conditions even if a single parameter, such as the wind speed, is used as a criterion. These differences range from the speed averaging period (mean or gust) and its probability of exceedance (frequency of occurrence) to the evaluation of its magnitude (experimental or computational).

Table 6.1 shows the traditional Beaufort scale used in ship navigation in a modified version applicable to land regions and for heights representative of pedestrians. This table provides an idea of the mechanical effects of wind of different speeds on the human body. Physiological effects are more complex since they depend on additional factors and their interactions.

A simple rule of thumb has been provided by Wise (1970) and Pendwarden (1973). This is based on mean speeds (V) assuming the following effects:

• $V = 5$ m/s or 18 km/h	Onset of discomfort
• $V = 10$ m/s or 36 km/h	Definitely unpleasant
• $V = 20$ m/s or 72 km/h	Dangerous

Conditions for pedestrians are considered acceptable if $V > 5$ *m/s less than 20 % of the time* (Penwarden and Wise 1975).

Recognizing the importance of frequency of occurrence along with the magnitude of wind speeds, Figs. 6.6, 6.7, and 6.8 provide threshold mean wind speeds for various types of activity as functions of the average annual number of storm

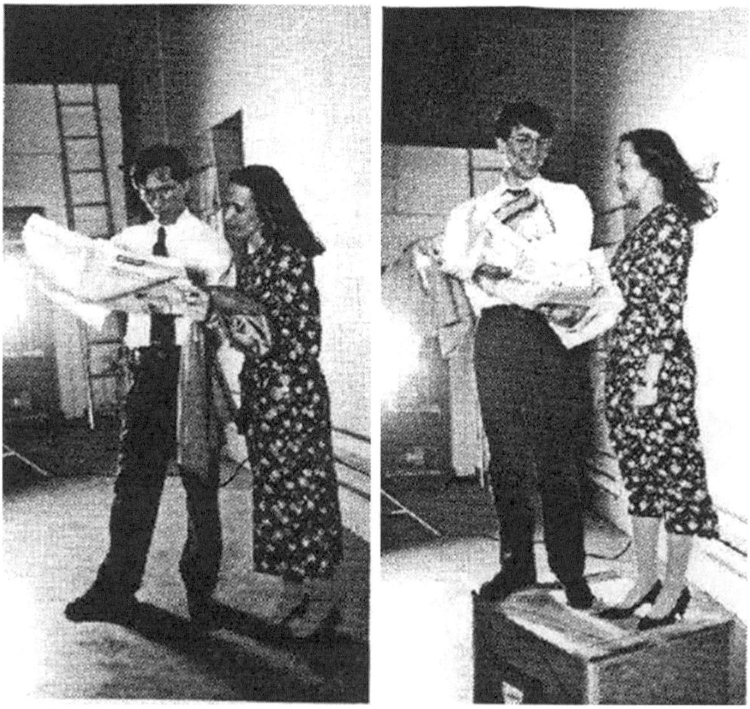

Fig. 6.6 Wind tunnel exposure of people at 10–15 km/h winds

occurrences. Naturally the mean wind speed threshold level drops significantly as the yearly average number of occurrences increases.

Wind tunnel experiments and observations of pedestrian performance suggest that $\kappa = 3$ is the most appropriate value. Figure 6.9 shows acceptance criteria for wind speeds for various annual frequencies of occurrence proposed by Isyumov and Davenport (1975). Note that these criteria are different from previous criteria in that, instead of specifying a wind speed for various activities, frequencies of occurrence are specified for different wind speeds. Murakami et al. (1986) produced the wind comfort criteria described in Table 6.2.

Melbourne (1978) has produced separate criteria based on mean and gust speeds. He proposed their application only for daylight hours and on the assumption that the max 2-s gust speed will be roughly twice as large as the mean speed, he produced the curves shown in Fig. 6.10. These curves identify threshold wind speed criteria for different types of activity similar to those shown in Table 6.2. Criteria for dangerous wind conditions were also specified. Such conditions are particularly important for cities with harsh winter conditions where icy sidewalks become source of frequent accidents when combined with high winds. Several cases of this nature have been reported, most involving accidents happened on elderly people. Liability issues are also interesting for such cases and courts have always a hard time dealing with them.

Fig. 6.7 Wind tunnel exposure of people at 20 (*left*) and 40 (*right*) km/h winds

On the basis of experience over a number of projects and wind tunnel studies, it has been concluded that Melbourne's criteria are on the strict side, i.e., if prevailing conditions abide by the prescribed limits, most sets of other criteria available in the literature or included in ordinances of various municipalities will be satisfied. Consequently, these criteria can be used as upper limits for pedestrian-level winds and, in this regard, are indeed valuable.

6.3.1 Wind Ordinances in Major Cities

There is great variation regarding wind ordinances in various cities around the world. In some cases, specific legislation has passed, and new building permits are not provided until the developers/owners demonstrate that the project will not generate dangerous or even uncomfortable and undesirable pedestrian-level wind conditions. In other cases, this is expected to happen as part of assumed good engineering and architectural practice. In general, the following points can be made:

Fig. 6.8 Wind tunnel exposure of people at 70 km/h winds

- Most major cities (Montreal, Toronto, Sydney, etc.) have some guidelines addressing the problem at the approval stage for new construction projects.
- San Francisco has adopted a very strict wind ordinance; they use $v^e = 42$ km/h with $P\ (>V^e) = 0.01\ \%$ as safety criterion; this is significantly lower than that proposed in most of the current literature.
- New York has strict air pollution standards, which tend to work against guidelines for the pedestrian wind environment; only 30 % of new developments have to go through a review process.
- The Boston Planning Department specifies that a wind tunnel study is required to assess wind environmental conditions near new developments for the following cases:

  (i) For any new building taller than 30 m (100 ft) and at least two times taller than its adjacent buildings
  (ii) For other buildings in special circumstances

As it is always the case with any adoption of code provisions or changes, passing legislation regarding pedestrian wind conditions is always problematic. It is worth mentioning that after several years of intense efforts by several experts, architects and engineers, a new wind ordinance has been approved in the Netherlands. It

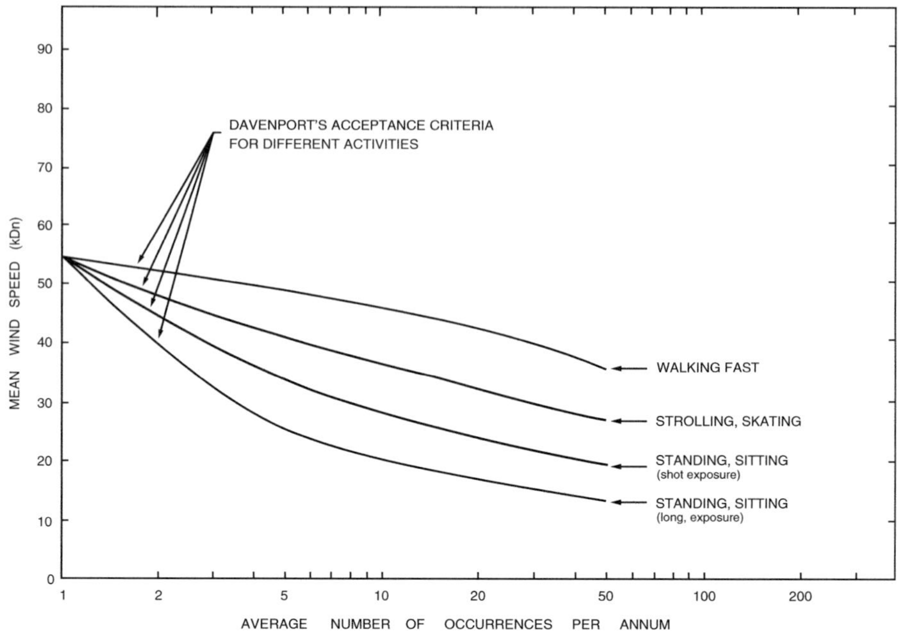

Fig. 6.9 Acceptance criteria for wind speeds for various annual frequencies of occurrence (After Isyumov and Davenport (1975))

Table 6.2 Wind environment criteria of Murakami et al. (1986)

Activity	Probability of $\hat{u} = 36$ kph	Exceedance $\hat{u} = 54$ kph	(P(>u)) $\hat{u} = 72$ kph
Long-term and short-term stationary exposure	0.10	0.008	0.0008
Strolling	0.22	0.036	0.006
Walking	0.35	0.07	0.015

represents the first standard on wind nuisance and wind danger in the world (NEN 2006a, b), and it is based on research by Verkaik (2000, 2006), Willemsen and Wisse (2002, 2007), Wisse and Willemsen (2003), and Wisse et al. (2007), among others. It was incited by the wide range in procedures and criteria being used by universities, research institutes, and consultancy companies across the country, sometimes yielding very different outcomes for the same building or urban area. In particular, the standard presents a uniform approach for each of the three main components of a wind comfort or wind danger study: the statistical meteorological data (including exposure corrections), the transformation of these data to the building site, and the comfort and danger criteria. A remarkable feature of the standard is that it explicitly allows the user to choose between wind tunnel modeling and CFD to determine the design-related contribution, as part of the transformation of wind statistics to the building site. Case studies applying this new

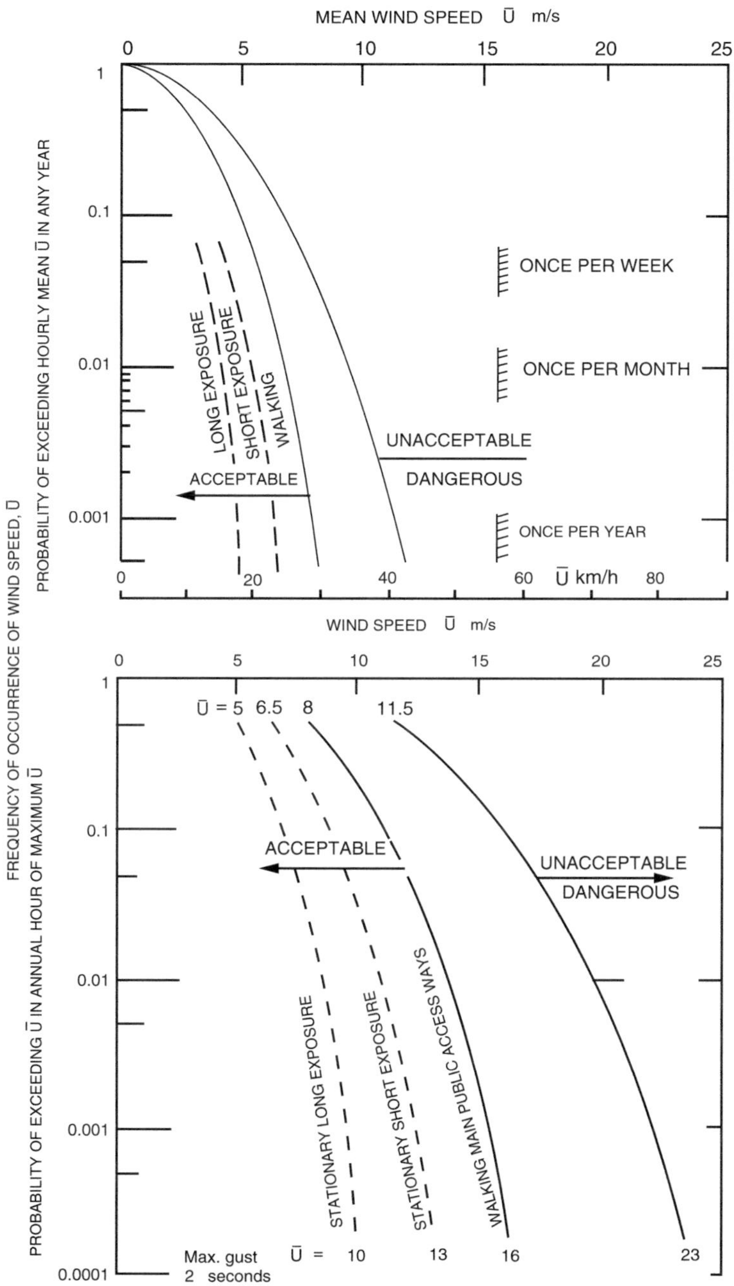

Fig. 6.10 Probability distributions of Melbourne's criteria for environmental wind conditions for daylight hours for a turbulence intensity of 30 % and $\hat{u} = 2u$ (After Melbourne (1978))

standard for different urban areas can be found in Blocken and Persoon (2009), Blocken et al. (2012), Janssen et al. (2013), and Montazeri et al. (2013).

6.4 Experimental Procedure: Wind Tunnel Approach

6.4.1 General

The testing of scale models in a boundary layer wind tunnel capable of simulating the mean velocity profile and turbulence, as well as the spectra of the natural wind, has been shown to be a very effective method of prediction by comparison with respective full-scale data. The wind tunnel model typically includes all buildings in the surrounding landscape; thus, their effect is automatically included. Both existing conditions and those with the new building(s) in place can be readily measured, thus allowing the impact of the new building(s) to be identified. Furthermore, the effects of changes to the building itself, or to landscaping, can also be studied, particularly where undesirable wind conditions are found.

A typical setup of a wind tunnel model in a boundary-layer wind tunnel is illustrated in Fig. 6.11. The building itself and the model of its surroundings are mounted on the wind tunnel turntable, which can be rotated to allow various wind directions to be simulated. Typical model scales for large buildings are in the range from 1:200 to 1:500. Larger scales have been used for smaller buildings. The model of surroundings enables the complex flows created by other buildings near the study

Fig. 6.11 Typical wind tunnel setup for a pedestrian wind assessment study in the atmospheric boundary layer wind tunnel at Concordia University, Montreal

building to be automatically included in the tests. However, it is also essential to create a proper simulation of the natural wind approaching the modeled area. The requirements for modeling the natural wind in a wind tunnel are described in the ASCE Manual of Practice (1999), as well as in the ASCE Standard 49-12, ASCE (2012). In typical wind tunnel tests, the airflow speed above the boundary layer is in the range 10–30 m/s. The process followed in the experimental approach consists of the following steps:

1. Meteorological records
2. Wind tunnel testing
3. Combination of (1) and (2)
4. Comparison with comfort criteria
5. Remedial measures

Details of this process will be presented in the following case study.

6.4.2 Case Study

This is an actual study in Montreal, but names have been withheld and sample results presented are also coming from another unnamed study for reasons of confidentiality. The study was conducted in the Building Aerodynamics Laboratory of Concordia University using a 1:500 model of the buildings and their surroundings. Wind environmental conditions at the sidewalks around the new buildings were assessed in terms of peak (gust) and mean speeds. The following configurations were considered in the study:

1. Present conditions
2. Proposed development

The architects of the project provided the details of the design, and the building models were placed on a maquette with the surroundings on a full-scale equivalent radius of 300 m. Environmental wind conditions on sidewalks around the building were assessed for the critical southwesterly and northeasterly wind directions – see Fig. 6.5. Measurements were obtained at the locations shown in Figs. 6.12 and 6.13. The 1:500-scaled model of the proposed development and the surroundings was mounted on the turntable of the boundary layer wind tunnel. Wind velocities were measured at 31 ground-level locations, but results for only six points are shown herein. Five of the sampling points (3, 5, 6, 7, and 8) were located on S street and one point (15) was located in the courtyard.

Pedestrian wind data were obtained with a hot film sensor located at a full-scale equivalent height of 2 m above the ground. The reference mean velocity in the wind tunnel at a full-scale equivalent height of 300 m, U_{300}, was also measured. The ratio $\hat{U}_{local}/U_{300}$ was determined for each wind direction, where $\hat{U}_{local}$ is the gust velocity at ground level. The test results were then combined with meteorological data

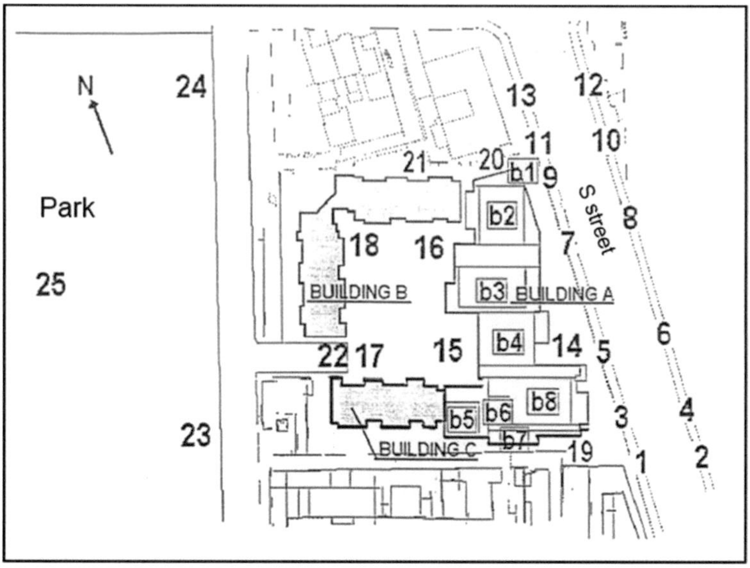

Fig. 6.12 Plan view of proposed development showing measurement locations

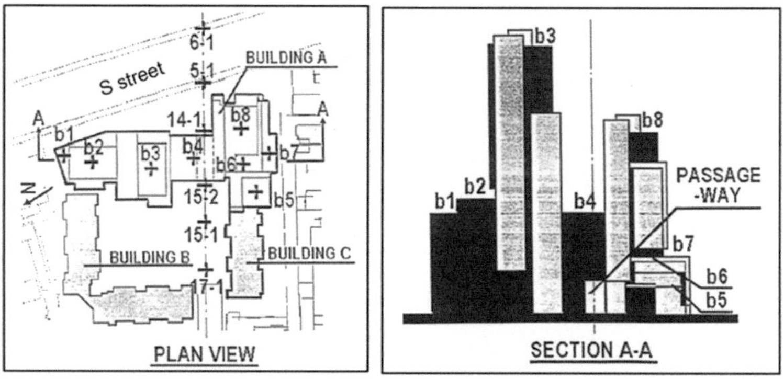

Fig. 6.13 Plan and elevation view of proposed development showing measurement locations near and on buildings

recorded at the P.E. Trudeau (Dorval) International Airport for a 10-year period in order to predict wind speeds with a particular return period for each test location. The Trudeau data, consisting of mean hourly wind speeds obtained in open country terrain at a height of 10 m, have been used to derive probability distributions of wind speed and direction for Montreal. These wind speeds were adjusted to an urban exposure at a height of 300 m using the following formulas:

$$U_{300,\text{open country}} = U_{10,\text{Dorval}} \left(\frac{300}{10}\right)^{0.15}$$

and

$$U_{300,\text{urban}} = U_{300,\text{open country}} \left(\frac{300}{450}\right)^{0.34}$$

since $U_{450,\text{urban}} = U_{300,\text{open country}}$. Note that the exposure category characterizes the terrain several kilometers upwind of the site.

Probability distributions of wind speed and direction for a suburban exposure at a height of 300 m similar to that presented in Fig. 6.5 have been evaluated for summer and winter, respectively. These distributions, which are based on data for daylight hours (07:00–19:00), indicate that strong winds occur most often from W, WSW, and SW directions and occur more frequently during the winter than in the summer. Data also show that the probability of strong winds from the NE is relatively high. The probability of exceedance [$P(>)$] corresponding to once per month and once per year are approximately 10^{-2} and 10^{-3}, respectively. For instance, a westerly wind at a height of 300 m is expected to exceed 48 kph approximately once per month during the winter. Although the proposed development will be highly exposed to winds from the east quadrant, the probability of strong winds from the east is relatively low. It should be noted that the influence of Mount Royal on Montreal's wind climate is not evident in the meteorological data used, since the probability curves were derived from data obtained at Trudeau (Dorval) Airport. It is anticipated that the presence of Mount Royal should reduce the probability of strong winds at the site for westerly and southwesterly winds. Therefore, the results presented in this study are expected to be generally conservative.

Melbourne's criteria shown in Table 6.3 have been used for the characterization of wind comfort conditions at the pedestrian level. The peak wind speed acceptable for walking was reduced to 32 kph for the winter months to take into account the effect of temperature on pedestrian comfort. As suggested in ASCE (2004), the summer wind speed should be reduced by one Beaufort number for every 20 °C reduction in temperature. The second criterion indicates that wind conditions are hazardous if the gust velocity exceeds 83 kph more than once per year. At this wind speed, people can be blown over.

Table 6.3 Melbourne's pedestrian comfort criteria (Melbourne 1978)

Wind condition	$\hat{U}_{local}$	$P(>\hat{U}_{local})$
Acceptable for walking		
Summer	48 kph	0.01 (once per month)
Winter	32 kph	0.01 (once per month)
Hazardous	83 kph	0.001 (once per year)

6.4.3 Sample Test Results

Data obtained under the current conditions and with the proposed development are plotted in polar form in Fig. 6.14 for a typical set of four measurement points. As suggested in Westbury et al. (2002), the data are plotted in terms of the velocity pressure ratio $(\hat{U}_{local}/U_{300})^2$. Melbourne's wind environment criteria for winter conditions have been included on the diagrams, in terms of curves determined by incorporating the probability distributions shown in Fig. 6.10. These curves show the limiting wind pressure values for walking comfort and for extreme hazard.

Figure 6.14 shows whether the proposed development does adversely affect the wind climate at different locations. For instance, the building causes an increase in gust speed at locations 6 and 8 for westerly and southwesterly winds and at location 6 for northeasterly winds. For these cases, Melbourne's comfort criteria were exceeded. However, it should be noted that the proposed development will improve wind conditions at all points for at least some wind directions. It can also be noted that at location 7, wind conditions were improved for all of the tested wind directions. The results of the study indicate that the proposed buildings will have relatively little adverse effect on the pedestrian wind environment. Winter winds, which are presented in Table 6.3, are more critical than those occurring during the summer. For the proposed development, the maximum wind gust expected to occur once per month in the winter is approximately 43 kph. Under current conditions (i.e., without the new building), the peak monthly gusts in the winter are approximately 33 kph at the same location.

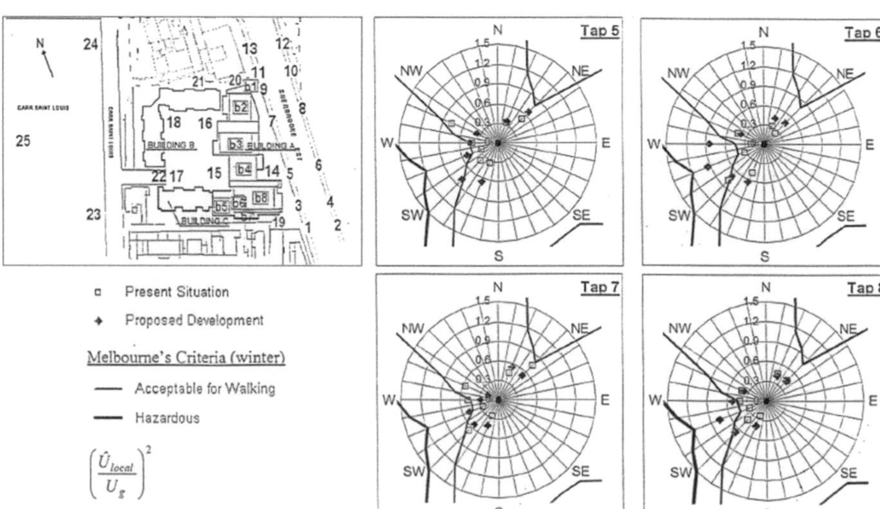

Fig. 6.14 Measurement of velocity pressure ratio at locations *5*, *6*, *7*, and *8*

6.4.4 Comparison with Montreal's Wind Criteria

The city's wind comfort criteria, specified in Article 39 of the *Règlements refondus de la ville de Montréal*, refer to *mean* wind speeds rather than *gust* speeds. The critical mean wind speeds, U_{local}, for winter and summer are 14.4 kph (4 m/s) and 21.6 kph (6 m/s), respectively, and the maximum acceptable probabilities of exceeding these values are as follows:

Location	Probability of exceedance of U_{local}
Main streets	15 %
Secondary streets	25 %
Parks	10 %

As previously mentioned, the wind tunnel data are expressed in terms of the local peak velocity pressure ratio, which is given as $(\hat{U}_{local}/U_{300})^2$, where $\hat{U}_{local}$ is the local gust velocity and U_{300} is the mean wind speed measured at the reference full-scale height of 300 m for an urban exposure. Using these data, the local mean velocity pressure ratio $(U_{local}/U_{300})^2$ is obtained by assuming that the peak velocity is two times the mean value, U_{local} (see Art. 38.5). Given the probability of exceedance of U_{300}, the probability of exceedance of any local mean wind speed – in this case, 14.4 kph (4 m/s) for winter conditions – can be determined for each wind direction. The total probability of exceedance is obtained by simply adding the probabilities determined for each wind direction.

For example, the gust velocity pressure ratio obtained at the assumed windiest location 6 with the proposed development for a west-southwesterly wind is 0.935 (see Fig. 6.14). Taking the square root and dividing by 2 gives the local mean velocity ratio of $U_{local}/U_{300} = 0.48$. Setting U_{local} equal to 14.4 kph gives the critical reference velocity of $U_{300} = 30$ kph for this wind direction. Based on wind speed data from Trudeau (Dorval) Airport, the probability of exceedance of this wind speed during daylight hours (07:00–19:00) in the winter months (November–April) is approximately 0.036 (3.6 %). Following this procedure for the other tested wind directions gives a total probability of occurrence of 0.0837 (8.37 %). For the wind directions not tested, the probability of $U_{local} > 14.4$ kph is expected to be very small – of the order of 1 %. Thus, the total probability of $U_{local} > 14.4$ kph at location 6 is approximately 9.37 %, as shown in Table 6.4. Consequently, the Montreal winter criterion for wind comfort on main streets ($U_{local} > 14.4$ kph less than 15 % of the time) is satisfied at location 6.

Furthermore, since location 6 is the windiest point in the project area, the Montreal standard is satisfied at all other locations and for all configurations tested in the study. This includes locations in an adjacent park, tested previously, even though the criterion is stricter for park areas ($U_{local} > 14.4$ kph less than 10 % of the time).

Table 6.4 Comparison with Montreal's wind criteria

Wind direction	$(\hat{U}_{local}/U_{300})^2$	$\hat{U}_{local}/U_{300}$	U_{local}/U_{300}	U_{300} (kph) at $U_{local} = 14.4$ kph	Probability $U_{local} > 14.4$ kph
NNE	0.421	0.649	0.32	45.0	0.0016
NE	0.456	0.675	0.34	42.4	0.0040
SSW	0.643	0.802	0.40	36.0	0.0008
SW	0.701	0.837	0.42	34.3	0.00095
WSW	0.935	0.967	0.48	30.0	0.0360
W	0.834	0.913	0.46	31.3	0.0300
WNW	0.381	0.617	0.31	46.3	0.0018
Other directions					0.0100
				Total	**0.0937**

Expected probabilities of exceeding the Montreal winter criterion ($U_{local} > 14.4$ kph) at location 6 due to the proposed development

The preceding example illustrates that strong winds at ground level are not likely to occur for a given wind direction unless $(\hat{U}_{local}/U_{300})^2$ is large, and the probability of strong winds from that direction is also large. Regarding the wind comfort criterion for the summer months, analysis of the data shows that the criterion will also be met at all locations. Due to the larger acceptable wind speed (21.6 kph) and the reduction in the frequency of high winds for the summer months, the probability of exceeding the summer criterion is less than that for the winter criterion at each location.

With respect to the city's criteria concerning hazardous wind conditions (*Art. 40, Règlements refondus de la ville de Montréal*), the wind conditions around the proposed development were found to be acceptable. The wind tunnel results show that predicted gust speeds with a probability of exceedance of 0.01 (once per month) are well below the criterion of 72 kph (20 m/s) at all measurement locations. For all design options, the maximum monthly gust in the winter is expected to be approximately 43 kph at locations 6 (on S street) and 15 (on passageway). Finally, it should be recalled that the anticipated wind speeds are based on statistical expectations and actual wind conditions during a particular storm may be different. Future building developments in the surrounding area may also affect the pedestrian wind environment, but this has not been considered in the present study.

6.5 Computational Procedure: CWE

In Computational Wind Engineering (CWE), the computer simulation essentially replaces the physical simulation in the boundary layer wind tunnel, at least in principle. Computational fluid dynamics (CFD) methods involve very large amounts of computation even for relatively simple problems, and their accuracy is often difficult to assess when applied to a new problem where prior experimental verification has not been done. Castro and Graham (1999), Stathopoulos (1997; 2002) and Blocken (2014) summarized the concerns expressed with respect to these issues. However, there have been cases for which the application of CFD methodologies appears to give somewhat satisfactory responses. These are cases requiring the determination of mean flow conditions and pressures, i.e., those related primarily with environmental issues. Typical problems of this category include but are not limited to pedestrian-level winds, snow dispersion and accumulation, dispersion of pollutants in the near-building and/or urban environment, ventilation, and the like. There is increasing evidence that for such problems CFD-based techniques may provide adequate responses – see Stathopoulos (2002) and Blocken (2014).

Pedestrian-level winds can be described quite adequately in terms of mean velocities in the presence and absence of a new building within a specific urban environment. Although it can be argued that pedestrians are mostly affected by gust effects and mean wind speeds may not be sufficient to produce satisfactory results, the fact remains that several major cities require only the satisfaction of certain mean (sustainable) speeds with a specified probability of exceedance. A number of computational studies for the evaluation of pedestrian-level winds and the comparison of their results with respective experimental data are encouraging. The process of comparison between computational and experimental results has already been challenged and appears problematic on its own. For instance, is it more meaningful to carry out point-by-point comparisons or does it make more sense to examine pedestrian-level wind speeds affecting a particular zone or area of influence for a specific activity within the urban environment? Furthermore, and after due consideration to the fact that pedestrian-level wind speeds measured in the proximity of buildings, i.e., in areas of high turbulence, are not very accurate, it may be conceivable that "errors" in the results might be better described in terms of their impact on design decisions. Clearly, this may be more reasonable, at least in the context of engineering perspective.

6.5.1 Case Studies

Bottema (1993) has attempted the evaluation of pedestrian-level wind conditions in the vicinity of an isolated building by using the CWE approach and a simple turbulence model but with only limited success. Studies published by Stathopoulos and Baskaran (1996) have demonstrated that by using a simple version of k-ε model, one may obtain computational results very comparable with corresponding values originating from respective wind tunnel studies. The comparison of computed and measured velocity ratios at 2 m above the ground level in the presence and absence of a proposed building is generally satisfactory with a maximum discrepancy of the order of 30 %. It is noteworthy that the maximum discrepancies between the experimental and numerical data appear in highly complex recirculating flow regions, for which neither the measured nor the computed values can be considered very accurate.

This type of data response has been found and confirmed by several authors. For instance, Timofeyef (1998) evaluated the wind flow around a five-storey-high development in Kazakhstan and produced full-scale results, wind tunnel data, and numerical results by using the discrete vortex method (two-dimensional flow). Surprisingly enough, computational results compare better with corresponding full-scale data than the latter with wind tunnel results. This means that this rather crude computational approach provides more representative results than wind tunnel testing, at least in this particular case. In a cooperative study between Portugal and Canada, Ferreira et al. (1999) produced wind flow around a group of low-rise buildings (Expo' 98 – Lisbon). Both wind tunnel and field data were compared with numerical results obtained with the standard formulation and the renormalization group (RNG) extension of the k-ε turbulence model. By and large the comparisons are satisfactory, at least for engineering design purposes.

Hu and Wang (2002) have also attempted the evaluation of street-level winds in a built-up area by using CFD. Computational velocity ratios obtained through the commercially available code "PHOENICS" agree reasonably well with the experimental results for surrounding blocks of similar height. However, the comparison results deteriorate when the surrounding blocks consist of buildings with different heights. Such trends are possibly due to the more significant interaction of horizontal separated flow with the downflow originating from the pressure difference at different heights.

Wind tunnel and CFD data have been compared for a complex development in central London, which includes a 40 m high apartment block adjacent to an 80 m tower block, and a number of other buildings – see Miles and Westbury (2002). This study forms part of a research program at BRE, which aims to assess whether CFD is "fit for purpose" for use during the design process and to determine the major sources of inaccuracy associated with user modeling decisions. The inlet mean velocity profile measured in the wind tunnel, representing an urban boundary layer, has been used in the CFD simulation. Turbulence kinetic energy and dissipation rate profiles have been derived using boundary layer parameters obtained from fitting the logarithmic law to the measured velocity profile, combined with expressions from Richards and Hoxey (1992). Wall boundary conditions were defined using a rough-wall turbulent wall function applied at the ground, and a smooth-wall function applied to the building envelopes. A number of turbulence models have been tested – see details in Westbury et al. (2002).

Finally, it is worth mentioning an intensive Japanese effort to compare CWE results in terms of pedestrian-level wind speeds with respective experimentally measured values in an actual environment – see Tominaga et al. (2004). The correspondence between the CFD and the experimental results is fairly good with the exception of the wake region.

6.6 Outdoor Comfort Issues

Outdoor human comfort in an urban climate depends on a wide range of weather and human factors. Studies have shown integrated effects of wind speed, air temperature, relative humidity, and solar radiation on the human perception, preference, and overall comfort in an urban environment. Some analysis of these issues has been presented in the ASCE SOA Report, ASCE (2004). Furthermore, the studies by Nicolopoulou et al. (2001) and Nicolopoulou and Lykoudis (2002) also address the influence of microclimatic characteristics in outdoor urban spaces and the comfort implications for the people using them. A significant characteristic is the psychological adaptation, which has also been addressed. An equivalent temperature has been defined and related to the outdoor human comfort by considering acclimatization and other biometeorological principles (Stathopoulos et al. 2004; Zacharias et al. 2001). However, the implications of this approach are far fetching and the overall assessment problems are still quite intriguing. Some basic ideas are presented here.

6.6.1 Temperature and Relative Humidity

Both can have a significant impact on a person's comfort, since sensation of comfort in cold conditions is linked to the heat balance of the human body, i.e., the balance of heat generated by metabolic processes and heat lost by conduction, convection, radiation, and evaporation. In convective and evaporative losses, the effects of temperature and humidity are closely linked with the wind conditions and cannot be treated in isolation from wind speed. This is why, for example, in the colder regions of Europe and North America, the wind chill equivalent temperature is used to provide a more meaningful description of how cold weather will really feel, rather than simply giving air temperature. The equivalent temperature is obtained by calculating the temperature in standard wind (set at 1.8 m/s = 4 mph) that would give the same rate of heat loss from the exposed skin at 33 °C as occurs in the actual wind and temperature conditions. Generally, in cold conditions, humidity is low and has little direct effect on thermal comfort, although there may be indirect effects, such as humidity changing the insulation value of clothing. In hot conditions, the human body needs to increase heat losses to maintain thermal comfort. This is largely achieved by reducing clothing and through sweating and the corresponding heat losses associated with the latent heat of evaporation. Since the efficiency of evaporation is decreased as the relative humidity of the air increases, the relative humidity becomes a much more important parameter in hot climates. Also, since the efficiency of evaporation is increased with wind speed, in cold climates it is often desirable to reduce wind speeds, but the opposite is sometimes the case in hot climates. The well-known humidex is an effective temperature, combining the temperature and humidity into one number to reflect the perceived temperature and to quantify human discomfort due to excessive heat and humidity. In general, almost everyone will feel uncomfortable when the humidex ranges from 40 to 45, and many types of labor must be restricted when the humidex is 46 and higher. The incorporation of relative humidity effects into the overall assessment of thermal comfort is discussed in Stathopoulos et al. (2004).

6.6.2 Solar Radiation

Any assessment of thermal comfort must account for the effects of sun/shade conditions. The angle of the sun, the amount of radiation absorbed by clouds, dust and particles in the atmosphere, and the sunlight absorbed and reflected by buildings need to be taken into account.

6.6.3 Precipitation

In heavy rain conditions, people are less likely to be outside, thus their wind and thermal comfort will usually be less critical compared with other microclimate factors. However, it may be of interest to evaluate how far under a sheltering canopy roof the precipitation will infiltrate and how often this will happen. Dampness of clothes may also be of interest because it will affect thermal comfort.

A working group of the International Society of Biometeorology has developed a new standardized universal thermal climate index (UTCI), which can also be used in the development of a criterion for human outdoor comfort (Hoppe 2002). An example of application of such an approach is shown in Stathopoulos et al. (2004). The dependence of the overall comfort is expressed on the basis of a group of survey respondents as a function of the difference of two equivalent temperatures: one based on the weather norm, $T_{e,n}$, and the other based on the actual outdoor conditions, $T_{e,a}$. Equivalent temperatures take into account the effect of relative humidity and solar radiation as well. It should be noted that $(T_{e,a} - T_{e,n})$ is the most influential factor on the overall comfort of the respondents. The study shows that (1) most comfortable conditions occur when the equivalent temperature difference is about 5 °C, which may be attributed to the preference of local residents for higher air temperature, as well as the temperature difference between an urban environment downtown and the airport; (2) lower comfort occurs with a negative temperature difference, or when the actual equivalent temperature is lower than the norm; and (3) if the temperature difference is beyond a certain limit, say greater than 10 °C, less comfortable (overall comfort < 1) outdoor conditions may be perceived, although more field data are necessary to confirm this observation. At present, it is still considered premature to draw a curve for a definite mathematical relationship of overall comfort and equivalent temperature difference.

6.7 Concluding Remarks

This chapter has dealt with pedestrian-level wind conditions, their origin, their experimental and, possibly, computational assessment in the urban environment, as well as with the criteria used for outside human comfort. Particular emphasis has been placed on the state of the art and the capabilities of Computational Wind Engineering to determine at least mean values of wind speeds in the vicinity of buildings in urban areas. An approach toward the establishment of an overall comfort index taking into account, in addition to wind speed, the temperature, and relative humidity in the urban area under consideration, was presented.

Acknowledgment The assistance of several graduate students in putting this chapter together and the financial support of the Natural Sciences and Engineering Research Council of Canada (NSERC) to the first author are gratefully acknowledged and highly appreciated.

References

ASCE (1999) Manual of practice No. 67 on wind tunnel studies of buildings and other structures. ASCE, Reston

ASCE (2004) Aerodynamics Committee, outdoor human comfort and its assessment, state of the art report. Task Committee on Outdoor Human Comfort, American Society of Civil Engineers, Reston

ASCE (2012) Wind tunnel testing for buildings and other structures, ASCE standard 49–12. American Society of Civil Engineers, Reston

Blocken B (2014) 50 years of computational wind engineering: past, present and future. J Wind Eng Ind Aerodyn 129:69–102

Blocken B, Persoon J (2009) Pedestrian wind comfort around a large football stadium in an urban environment: CFD simulation, validation and application of the new Dutch wind nuisance standard. J Wind Eng Ind Aerodyn 97(5–6):255–270

Blocken B, Carmeliet J, Stathopoulos T (2007) CFD evaluation of wind speed conditions in passages between parallel buildings – effect of wall-function roughness modifications for the atmospheric boundary layer flow. J Wind Eng Ind Aerodyn 95(9–11):941–962

Blocken B, Stathopoulos T, Carmeliet J (2008a) Wind environmental conditions in passages between two long narrow perpendicular buildings. J Aerosp Eng, ASCE 21(4):280–287

Blocken B, Moonen P, Stathopoulos T, Carmeliet J (2008b) A numerical study on the existence of the Venturi-effect in passages between perpendicular buildings. J Eng Mech, ASCE 134(12): 1021–1028

Blocken B, Janssen WD, van Hooff T (2012) CFD simulation for pedestrian wind comfort and wind safety in urban areas: general decision framework and case study for the Eindhoven University campus. Environ Model Software 30:15–34

Bottema M (1993)Wind climate and urban geometry. PhD dissertation, Technische Universitiet Eindhoven, Facultyeit Bouwkunde, Vakgroep Fago

Castro IP, Graham JMR (1999) Numerical wind engineering: the way ahead? Proc Inst Civ Eng Struct Build 134:275–277

Cochran LS (2004) Design features to change and/or ameliorate pedestrian wind conditions. In: Proceedings of the ASCE structures congress, Nashville, May 2004

Ferreira AD, Viegas DX, Sousa ACM (1999) Numerical and experimental study of the wind flow around a group of low-rise buildings. In: Proceedings of the 10th international conference on wind engineering, Copenhagen, 21–24 June 1999
Hoppe P (2002) Different aspects of assessing indoor and outdoor thermal comfort. Energy Build 34:661–665
Hu C-H, Wang F (2002) Personal communication
Isyumov N, Davenport AG (1975) The ground level wind environment in built-up areas. In: Proceedings of the 4th international conference on wind effects on buildings and structures, London
Janssen WD, Blocken B, van Hooff T (2013) Pedestrian wind comfort around buildings: comparison of wind comfort criteria based on whole-flow field data for a complex case study. Build Environ 59:547–562
Lawson TV, Penwarden AD (1975) Effect of wind on people in the vicinity of buildings. In: Proceedings of the 4th international conference on wind effects on buildings and structures, Heath-row, pp 605–622
Melbourne WH (1978) Criteria for environmental wind conditions. J Wind Eng Ind Aerodyn 3: 241–249
Miles SD, Westbury PS (2002) Assessing CFD as a tool for practical wind engineering applications. In: Proceedings of the 5th UK conference on wind engineering, University of Nottingham, Sept 2002
Montazeri H, Blocken B, Janssen WD, van Hooff T (2013) CFD evaluation of new second-skin facade concept for wind comfort on building balconies: case-study for the Park Tower in Antwerp. Build Environ 68:179–192
Murakami S, Iwasa Y, Morikawa Y (1986) Study on acceptable criteria for assessing wind environment on ground level based on residents' diaries. J Wind Eng Ind Aerodyn 24:1–18
NEN (2006a) Wind comfort and wind danger in the built environment, NEN 8100 (in Dutch) Dutch Standard
NEN (2006b). Application of mean hourly wind speed statistics for the Netherlands, NPR 6097:2006 (in Dutch) Dutch practice guideline
Nicolopoulou M, Lykoudis S (2002) Thermal comfort in open spaces: the human factor. In: Proceedings of the 6th Panhellenic congress of meteorology, climatologic and atmospheric physics, Ioannina, 25–28 Sept 2002
Nicolopoulou M, Baker N, Steemers K (2001) Thermal comfort in outdoor urban spaces: understanding the human parameter. Sol Energy 70(3):227–235
Penwarden AD (1973) Acceptable wind speeds in towns. Build Sci 8:259–267
Penwarden AD, Wise AFE (1975) Wind environment around buildings. Building Research Establishment Digest, Bracknell, Berkshire RG12 8FB
Richards PJ, Hoxey RP (1992) Computational and wind tunnel modeling of mean wind loads on the Silsoe structures building. J Wind Eng Ind Aerodyn 41–44:1641–1652
Stathopoulos T (2002) The numerical wind tunnel for industrial aerodynamics: real or virtual in the new millennium? Wind Struct 5(2–4):193–208
Stathopoulos T (1997) Computational wind engineering: past achievements and future challenges. J Wind Eng Ind Aerodyn 67–68:509–532
Stathopoulos T, Baskaran A (1996) Computer simulation of wind environmental conditions around buildings. Eng Struct 18(11):876–885
Stathopoulos T, Wu H, Zacharias J (2004) Outdoor human comfort in an urban climate. Build Environ 39(3):297–305
Timofeyef N (1998) Numerical study of wind mode of a territory development. In: Proceedings of the 2nd East European conference on wind engineering, Prague, 7–11 Sept 1998
Tominaga Y, Mochida A, Shirasawa T, Yoshie R, Kataoka H, Harimoto K, Nozu T (2004) Cross comparisons of CFD results of wind environment at pedestrian level around a high-rise building and within a building complex. J Asian Architect Build Eng 3:63–70

Venturi GB (1799) Experimental enquiries concerning the principle of the lateral communication of motion in fluids: applied to the explanation of various hydraulic phenomena. Translated from the French by W. Nicholson, 1st English Ed., J. Taylor, Architectural Library, High-Holborn, London

Verkaik JW (2000) Evaluation of two gustiness models for exposure correction calculations. J Appl Meteorol 39(9):1613–1626

Verkaik JW (2006) On wind and roughness over land. PhD thesis, Wageningen University, Wageningen, 123 pp

Westbury PS, Miles SD, Stathopoulos T (2002) CFD application on the evaluation of pedestrian-level winds. Workshop on impact of wind and storm on city life and built environment, cost action C14, CSTB, Nantes, 3–4 June 2002

Willemsen E, Wisse JA (2002) Accuracy of assessment of wind speed in the built environment. J Wind Eng Ind Aerodyn 90:1183–1190

Willemsen E, Wisse JA (2007) Design for wind comfort in the Netherlands: procedures, criteria and open research issues. J Wind Eng Ind Aerodyn 95(9–11):1541–1550

Wise AFE (1970) Wind effects due to groups of buildings. In: Proceedings of the Royal Society symposium architectural aerodynamics, session 3, Effect of buildings on the local wind, London, February 1970, pp 26–27

Wisse JA, Willemsen E (2003) Standardization of wind comfort evaluation in the Netherlands. In: Proceedings of the 11th international conference on wind engineering (11ICWE), Lubbock

Wisse JA, Verkaik JW, Willemsen E (2007) Climatology aspects of a wind comfort code. In: Proceedings of the 12th international conference on wind engineering (12ICWE), Cairns

Zacharias J, Stathopoulos T, Wu H (2001) Microclimate and downtown open space activity. Environ Behav 33(2):296–315

Chapter 7
Wind-Induced Dispersion of Pollutants in the Urban Environment

Ted Stathopoulos and B. Hajra

Abstract Predicting air and pollutant flow around buildings in an urban environment is a very complex problem affecting building design and performance. This chapter presents some of the new developments in this field, as far as the assessment of pollutant concentrations is concerned and the evolving design guidelines in this area. Particular emphasis is placed on the results of wind tunnel studies to assess the influence of adjacent buildings and rooftop structures on near-field pollutant dispersion by considering various parameters, such as stack height, exhaust momentum and spacing between buildings. A general discussion of the various ASHRAE models, as well as comparisons with wind tunnel results for a few adjacent building configurations, is presented. Application of ADMS, a Gaussian-based dispersion model, on near-field pollutant dispersion is also discussed. Comparisons for computational fluid dynamics (CFD) results and wind tunnel data for a particular case are made. The limitations of ASHRAE and CFD models to predict realistic dilutions for particular building configurations, besides suggestions to improve them, are discussed. Guidelines regarding appropriate stack and intake locations to avoid plume reingestion are also presented.

Keywords ASHRAE • Building • Dilution • Wind tunnel • CFD

T. Stathopoulos (✉)
Department of Building, Civil and Environmental Engineering, Concordia University, Montreal, QC H3G 1M8, Canada
e-mail: statho@bcee.concordia.ca

B. Hajra
Department of Building, Civil and Environmental Engineering, Concordia University, Montreal, QC H3G 1M8, Canada

Present address: International Hurricane Research Centre, Florida International University, Miami, FL, USA
e-mail: bhajra@fiu.edu

Y. Tamura, R. Yoshie (eds.), *Advanced Environmental Wind Engineering*,
DOI 10.1007/978-4-431-55912-2_7

7.1 Introduction

This chapter focuses on the effects of wind-induced pollutant dispersion from rooftop emissions in the built environment through various experimental and analytical studies, carried out in the last decade. The experimental results mainly originate from wind tunnel measurements, while analytical techniques pertain to ASHRAE and ADMS formulations. Additionally, a brief discussion on the application of computational fluid dynamics (CFD) models is presented. Current standards for building ventilation systems recommend that rooftop stacks from industrial, laboratory or hospital buildings be designed such that their emissions do not contaminate fresh air intakes of the emitting building or nearby buildings. Unfortunately, the state of the art has not been sufficiently advanced to allow building engineers to apply appropriate design criteria to avoid this problem for new construction or to help alleviate it for existing buildings. Several instances of potential health hazards due to pollutant reingestion from isolated buildings, as well as limited studies on adjacent building configurations, are available in the literature (e.g. Schulman and Scire 1991; Wilson et al. 1998). Recently, Hajra et al. (2011) performed a detailed wind tunnel study to assess the effects of upstream buildings on near-field pollutant concentrations and found that the height and across wind dimensions of the upstream building, as well as the spacing between buildings, were critical parameters in altering the plume geometry. More recently, Hajra and Stathopoulos (2012) showed that a taller downstream building prevents pollutants from dispersing, thereby increasing rooftop concentrations on the emitting building.

Besides wind tunnel studies, field measurements of tracer studies were also carried out in the past. For instance, field studies by Wilson and Lamb (1994) have shown that even with high exit velocities and moderately high stacks, pollutant concentrations may be unacceptably high at particular locations. Studies by Stathopoulos et al. (2004) carried out on two of the buildings located at Concordia University, Canada, have also shown that a taller upstream building, with effluents being released from a lower downstream building, may affect the leeward wall of the upstream building, as well as the roof of the emitting building. In general, several factors may account for the occasional poor performance of rooftop stacks. These factors include the location of the stack relative to regions of flow separation and flow reattachment, the presence of rooftop irregularities such as penthouses and high upstream turbulence. Relatively few studies have compared wind tunnel concentration data with field measurements for near-field diffusion cases (i.e. receptors within 50 m of a stack). This is one of the most difficult fluid modelling applications, since the plume characteristics may be sensitive to a number of local factors (building wake effects, the position of the stack relative to rooftop recirculation zones, stack Reynolds number, etc). On the other hand, for far-field applications, plume characteristics are much less sensitive to these factors. Higson et al. (1994) conducted field tracer gas experiments with a stack at varying distances upwind of a small building and found that the maximum concentrations were generally overestimated in the wind tunnel tests, while the minimum concentrations were underestimated. This suggests that the wind tunnel plume was narrower than the field plume due to the absence of large-scale turbulence in the wind tunnel. Several studies at Concordia University have evaluated the accuracy of wind tunnel dispersion measurements (Stathopoulos et al. 1999, 2004) and found generally good agreement between wind tunnel and field data. The wind tunnel concentration values were usually within a factor of 2 of the field values. In general, the accuracy of the wind tunnel generally increases as stack-receptor distance increases.

Various semiempirical models have been developed by ASHRAE for estimating near-field dilution of plumes emitted from rooftop stacks for open fetch situations. The models can be used to assess the minimum stack height to avoid plume reingestion (geometric design method) and Gaussian plume equations to assess plume dilutions on rooftop receptors. These models include ASHRAE (1997, 1999, 2003, 2007, 2011). ASHRAE (1997, 1999) are mainly based on the works of Halitsky (1963), while the 2003 and 2007 versions are based on the efforts of Wilson (1979). The more recent version of ASHRAE 2011 has been developed primarily by Petersen et al. (2004). The accuracy of these models has been evaluated by various researchers, and with the exception of ASHRAE 2011, most of these models were found to be overly conservative for isolated buildings (Stathopoulos et al. 2004; Hajra et al. 2011, 2013). Additionally, these models were found to be incapable of simulating the effects of rooftop structure and adjacent buildings (Stathopoulos et al. 2008; Gupta et al. 2012). However, ASHRAE 2011 predictions were found to be reasonably accurate for an isolated building with low exhaust speeds. Furthermore, the 2011 version also provides guidelines to determine dilutions on the sidewalls of an emitting building, as opposed to previous versions that were only restricted to rooftop dilutions on the source (Hajra et al. 2013). Dispersion models developed by the Environmental Protection Agency (EPA) are mainly suited for far-field pollutant dispersion problems, as they cannot model the turbulence caused by buildings and structures in the near vicinity of the source (Stathopoulos et al. 2008). CFD has been used extensively in the last few decades to model pollutant dispersion in the urban environment (Tominaga and Stathopoulos 2007). Unfortunately, CFD simulations cannot accurately predict the air and pollutant flow characteristics around buildings, making it necessary to carry out additional experimental studies for realistic urban layouts, in order to improve CFD (Blocken et al. 2011).

7.2 ASHRAE Dispersion Models

The various versions of ASHRAE described in this section include 1997, 1999, 2003, 2007 and 2011. The dilution equations in 1997 and 1999 versions are essentially non-Gaussian, while the 2003, 2007 and 2011 versions are semi-Gaussian models. It is worth mentioning that the ASHRAE models use the geometric design method, which is common to all versions, and the equations for estimating dilutions, which are somewhat different for each version.

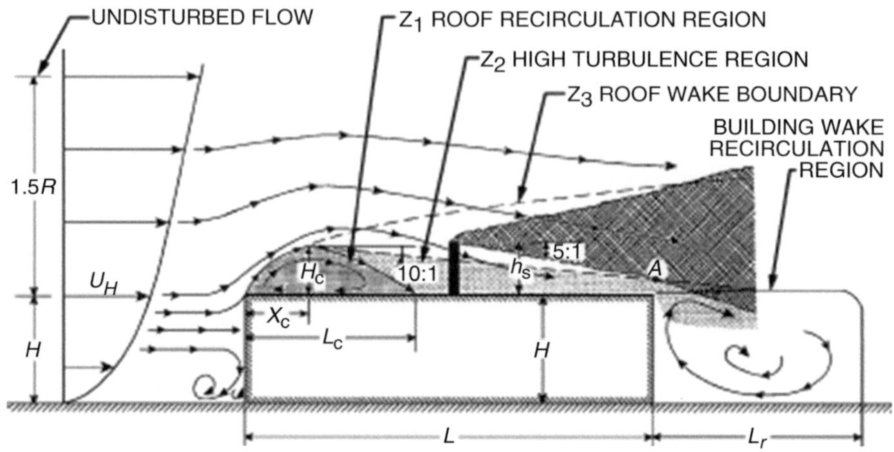

Fig. 7.1 Design procedure for required stack height to avoid contamination (Wilson 1979)

7.2.1 Geometric Design Method

The geometric design method provided in all versions of ASHRAE (1997, 1999, 2003, 2007, 2011) has remained unchanged. This method is used to estimate the minimum stack height to avoid plume entrainment in the flow recirculation zones of a building and its rooftop structures. Dimensions of the recirculation zones are expressed in terms of the building dimensions (Fig. 7.1):

$$L_r = {B_s}^{0.67}{B_L}^{0.33} \tag{7.1}$$

where B_s is the smaller of upwind building height or width and B_L is the larger of these dimensions. The dimensions of flow recirculation zones that form on the building and rooftop structures are

$$H_c = 0.22L_r \tag{7.2}$$

$$X_c = 0.5L_r \tag{7.3}$$

$$L_c = 0.9L_r \tag{7.4}$$

where H_c is the maximum height of the roof recirculation zone, X_c is the distance from the leading edge to H_c, L_c is the length of the roof recirculation zone and L_r is the length of the building wake zone.

The design method assumes that the boundary of the high turbulence region is defined by a line with a slope of 10:1 extending from the top of the leading edge separation bubble. The location of the plume relative to the recirculation zones is determined by taking into account plume rise due to exhaust momentum and assuming a conical plume with a slope of 5:1. It should be noted that the geometric design method is applicable for wind directions that are approximately normal to the windward wall of the building, which is the critical case for flat-roof buildings

(Saathoff et al. 2009). On the other hand, for oblique winds, building-generated turbulence is normally not significant in the central part of the roof (where stacks are usually located). For such winds, the building-generated turbulence is confined to the leading edges of the roof where the familiar delta-wing vortices are formed.

7.2.2 Gaussian Plume Equations

Gaussian plume equations have remained unchanged for 1997 and 1999 versions, while changes have been suggested to the 2003, 2007 and 2011 versions, as will be discussed further herein.

7.2.2.1 ASHRAE 1997/1999

Two different sets of equations for isolated building with flush vent and short stacks were part of the model. The dilution, defined as the ratio of the exhaust concentration (C_e) to receptor concentration (C_r), for buildings with short stack (less than 3 m), was based on wind tunnel experimental works of Halitsky (1963):

$$D_{\min} = \left[\alpha + 0.11(1 + 0.2\alpha)S/{A_e}^{0.5}\right]^2 \tag{7.5}$$

where:

$D_{\min}$ is the minimum dilution for a given wind speed.
S is the distance between a particular rooftop receptor and the stack (m).
α is a factor, which incorporates building shape and orientation, and M (generally $\alpha = 1$).
A_e is the exhaust area (m^2).

Similarly, the dilution equation for an isolated building with a flush vent was based on the works of Wilson and Lamb (1994), Wilson and Chui (1985, 1987) and Chui and Wilson (1988):

$$D_{\min} = \left[{D_o}^{0.5} + {D_s}^{0.5}\right]^2 \tag{7.6}$$

where:

D_o is the initial dilution caused by the turbulence in the exhaust jet.
D_s is the dilution caused by the atmospheric and building effects.

It may be noted that Eqs. 7.5 and 7.6 did not consider the wake recirculation length (Eq. 7.1) to estimate the dilution. Some earlier studies by Saathoff et al. (1998) and Stathopoulos et al. (1999) have also shown the highly conservative nature of ASHRAE 1997, by comparing the model to field concentration measurements carried out on some of the buildings at Concordia University. This led to the development of the ASHRAE 2003 model as will be also discussed further.

7.2.2.2 Gaussian Dilution Model (ASHRAE 2003, 2007, 2011)

The Gaussian dilution model recommended in ASHRAE 2003 is based on a series of experiments carried out in a water flume by Wilson et al. (1998). The model predicts worst-case dilution at roof level, D_r, assuming that the plume has a Gaussian (bell-shaped) concentration profile in both the vertical and horizontal directions. The effective height of the plume above the roof or rooftop structure is

$$h = h_s + h_r - h_d \tag{7.7}$$

where h_s is stack height, h_r is plume rise and h_d is the reduction in plume height due to entrainment into the stack wake during periods of strong winds. It should be noted that h_s is the height of the stack tip above the roof minus the height of rooftop obstacles (including their recirculation zones) that are in the path of the plume.

Plume rise, which is assumed to occur instantaneously, is calculated using the formula of Briggs (1984):

$$h_r = 3\beta d_e (V_e/U_H) \tag{7.8}$$

where d_e is the stack diameter, V_e is the exhaust velocity, U_H is the wind speed at building height and β is the stack capping factor. The value of β is 1 for uncapped stacks and 0 for capped stacks. The ratio of the exhaust velocity (V_e) to the wind velocity at the building height (U_H) is called exhaust momentum ratio (M). The formulations for plume rise estimate (see Eq. 7.7) has remained unchanged for the 2003 and 2007 versions. However, the 2011 version has implemented the following changes:

$$h_r = \min\{\beta h_x, \beta h_f\} \tag{7.9}$$

where:

h_x and h_f are estimated as

$$h_x = \frac{3{V_e}^2{d_e}^2 X}{4{\beta_j}^2{U_H}^2} \tag{7.10}$$

$$h_f = \frac{0.9\left[\left({V_e}^2{d_e}^2/4\right)\left(U_H/U_*\right)\right]^{0.5}}{\beta_j U_H} \tag{7.11}$$

where:

U_* is the friction velocity (m/s).
β_j is the jet entrainment coefficient calculated by

$$\beta_j = \frac{1}{3} + \frac{U_H}{V_e} \tag{7.12}$$

The logarithmic wind profile equation is

$$U_H/U_* = 2.5\ln(H/Z_o) \tag{7.13}$$

where:

Z_o is the surface roughness length (m).

As per ASHRAE 2003, the roof-level dilution for a plume at height h at a receptor distance X from the stack is given as

$$D_r = 4\frac{U_H}{V_e}\frac{\sigma_y}{d_e}\frac{\sigma_z}{d_e}\exp\left[\frac{h^2}{2\sigma_z^2}\right] \tag{7.14}$$

where U_H is the wind speed at the building height, d_e is stack diameter, V_e is the exhaust speed and σ_y and σ_z are the plume spreads in the horizontal and vertical directions, respectively. The equations for σ_y and σ_z are adjusted from a 60 min averaging time to a 2 min averaging time using the 0.2 power:

$$\frac{\sigma_y}{d_e} = 0.071\left(\frac{t_{avg}}{2.0}\right)^{0.2}\frac{X}{d_e} + \frac{\sigma_o}{d_e} \tag{7.15}$$

$$\frac{\sigma_z}{d_e} = 0.071\frac{X}{d_e} + \frac{\sigma_o}{d_e} \tag{7.16}$$

where t_{avg} is the concentration averaging time in minutes and σ_o is the initial source size that accounts for stack diameter and for dilution due to jet entrainment during plume rise.

It must be noted that Eqs. 7.15 and 7.16 remain unchanged in 2003 and 2007 versions. However, ASHRAE 2011 uses the formulations of Cimoreli et al. (2005) to estimate the spread parameters:

$$\sigma_y = \left(i_y{}^2X^2 + \sigma_o{}^2\right)^{0.5} \tag{7.17}$$

$$\sigma_z = \left(i_z{}^2X^2 + \sigma_o{}^2\right)^{0.5} \tag{7.18}$$

$$i_y = 0.75i_x \tag{7.19}$$

$$i_z = 0.5i_x \tag{7.20}$$

$$i_x = \left[0.24 + 0.096\log_{10}(Z_o) + 0.016(\log_{10}Z_o)^2\right][\ln(30/Z_o)/\ln(Z/Z_o)] \quad (7.21)$$

where:

i_x, i_y and i_z are the turbulence intensities in x, y and z directions.
σ_o is the initial source size and is set equal to $0.35d_e$ (m).
Z is the height of the building (m).

According to ASHRAE 2011, calculations must be carried out separately for Z_o, $0.5Z_o$ and $1.5Z_o$, and the lowest dilution must be considered for the design. Additionally, following the recommendation of Petersen et al. (2004), dilutions can be estimated on the sidewall of a building based on the dilution obtained on the nearest rooftop receptor, by increasing the latter by a factor of 2 (for conservative values). This is an important contribution of ASHRAE 2011 model, since previous versions can only be used to assess dilutions on the rooftop of the emitting building. The Gaussian dilution model (Eq. 7.14) should not be used when the plume height, h, is less than the maximum height of the roof recirculation zones that are in the path of the plume. The dilution equation used in ASHRAE 2007 and 2011 is

$$D_r = 4(U_H/V_e)(\sigma_y/d_e)(\sigma_z/d_e)\exp(\zeta^2/2\sigma_z^2) \quad (7.22)$$

where:

$$\begin{aligned}\zeta &= h - H_c\\ &= 0 \text{ if } h < H_c\end{aligned}$$

ζ is the vertical separation between 'h' and H_c.

ASHRAE 2011 also provides a detailed discussion on averaging time effects on near-field dilutions, which is a significant contribution since this was not part of previous versions. It is worth noting that ASHRAE 2011 considers the dilution estimates equivalent to 10–15 min averaging time, while the 2007 version assumes these dilutions to be equivalent to 2 min and considers it to be constant for longer averaging times. However, averaging time greatly influences the dispersion process especially at the micro-scale level, as discussed in Hajra et al. (2011). ASHRAE 2007 limits the value of $\zeta^2/2\sigma_z^2$ to 7, while the 2011 version does not impose any limits, which has led the latter to predict even more conservative estimates compared to ASHRAE 2007 for higher exhaust speeds ($M > 3$), as will be discussed further. Table 7.1 summarises the main features and contributors of each model.

Table 7.1 Summary of various ASHRAE dispersion models and their respective features

Model	[a]Based on the works of	Main features
ASHRAE 1997/1999	Halitsky (1963) Wilson (1979, 1982) Wilson and Lamb (1994) Wilson and Chui (1985, 1987) Chui and Wilson (1988)	Adopts a non-Gaussian approach Presents separate formulae for rooftop stacks and flush vents Assumes the calculated dilutions are for 10 min averaging time
ASHRAE 2003 and 2007	Wilson (1979, 1982) Wilson et al. (1998) Briggs (1984)	Limits $h^2/2\sigma_z^2$ to 5 for ASHRAE 2003 and 7 for ASHRAE 2007, close to the stack Considers σ_y and σ_z to be functions of exhaust diameter and receptor distance Assumes initial spread (σ_o) to be function of M Assumes dilution estimates from Eqs. 7.3 and 7.4 for 2 min averaging time and considers the dilution values to be constant for longer averaging times
ASHRAE 2011	Wilson (1979, 1982) Wilson et al. (1998) Cimoreli et al. (2005) Petersen et al. (2004)	Considers no limit for $h^2/2\sigma_z^2$ close to the stack Assumes σ_y and σ_z to be functions of turbulence intensities and receptor distance Assumes initial spread (σ_o) equal to $0.35d_e$ States explicitly that dilution estimates for ASHRAE 2011 are for 10–15 min averaging time

[a]Only main contributors are mentioned

7.3 Application of Dispersion Models to Near-Field Pollutant Dispersion Problems

Dispersion models are essentially computer-based models that can solve Gaussian equations to assess pollutant concentrations at a given rooftop/ground receptor. The Environmental Protection Agency (EPA), USA, has developed several models that can be applied to various situations (e.g. accidental release, release of pollutants from point or area sources, etc). Most of these models require inputs in the form of meteorological data, rate and type of pollutant releases and receptor locations, while the output is generated as concentrations at specific locations or contour plots. Only some of these models, such as Atmospheric Dispersion Modelling System (ADMS) and SCREEN, have been applied to near-field pollutant dispersion problems, because the majority of them cannot simulate the turbulence caused by buildings and only apply to far-field situations (Hajra et al. 2010). In fact, ADMS is based on the model of Hunt and Robins (1982) and has been validated using several field studies (e.g. Robins and McHugh 2001). However, some recent studies by Hajra et al. (2010) have shown that ADMS cannot model near-field pollutant dispersion, because it assumes a uniform concentration field in the wake of the building. Furthermore, ADMS is incapable of simulating the effect of a rooftop structure (RTS), which can greatly affect plume concentrations on the building roof. According to Riddle et al. (2004), 'such atmospheric dispersion packages are not able to assess the local effects of a complex of buildings on the flow field and turbulence, and whether gas will be drawn down amongst the buildings'. A detailed description of various EPA models, and their applications to near-field plume dispersion, is discussed in Stathopoulos et al. (2008).

7.4 Application of CFD Models to Near-Field Pollutant Dispersion Problems

CFD has been an emerging tool in the last few decades mainly because of increased cost associated with experimental work, coupled by the development of high-speed computers which can facilitate faster computing methods to solve the complex 'Navier-Stokes equations'. CFD mainly consists of Reynolds-averaged Navier-Stokes (RANS)-based models (steady and unsteady) and Lagrangian and large eddy simulations (LES). The main problem with the application of CFD for near-field dispersion problems is that they cannot accurately simulate the turbulence caused by buildings and structures (Tominaga and Stathopoulos 2011). In fact studies have also shown that CFD simulations are extremely sensitive to turbulent Schmidt number (Sct) variations (Tominaga and Stathopoulos 2007). According to ASHRAE 2011, 'Based on the current state of the art, CFD models should be used with extreme caution when modelling exhaust plumes from laboratory pollutant sources. Currently, CFD models can both over- and underpredict concentration levels by orders of magnitude, leading to potentially unsafe designs'. This explains the need to carry out additional experiments (wind tunnel/field) for realistic scenarios and utilise those results to improve CFD in the future. A detailed discussion of CFD is beyond the scope of this chapter. However, one may refer to Stathopoulos (1997) and Blocken et al. (2011) for additional information on the subject.

7.5 Selected Results from Recent Studies

In this section, the results from previous studies are classified in the following sections:

(a) Wind tunnel, ADMS and ASHRAE comparisons
(b) Wind tunnel and ASHRAE dilution results
(c) CFD, wind tunnel and ASHRAE comparisons

7.5.1 Wind Tunnel, ADMS and ASHRAE Comparisons

Figure 7.2a shows comparisons for wind tunnel data, ADMS and ASHRAE 2003 and 2007 for an isolated building without an RTS, with a centrally placed stack of 1 m and M = 3, in terms of normalised dilution. The building used in this study was 15 m high and 50 m square in plan; additional experimental details can also be found in Stathopoulos et al. (2008). Results show that ADMS predicts about five times higher dilutions than wind tunnel data at all receptors beyond 5 m from the stack. This is primarily because ADMS assumes a uniform concentration distribution in the wake of the building. ASHRAE 2003 and 2007 predict comparable dilutions at all receptors, although both models predict about ten times lower dilutions than wind tunnel data. This is because ASHRAE predicts lower plume rise, resulting in higher rooftop concentrations (lower dilutions) at all receptors. The trends are somewhat different for the same building with an RTS, as shown in Fig. 7.2b, where ASHRAE 2007 predicts about five times lower dilutions than wind tunnel data at all receptors due to reasons explained previously. However, ASHRAE 2003 predictions are about five times higher than wind tunnel data because the value of the exponential term in the 2003 version is limited to 5, while ASHRAE 2007 limits this value to 7 (Table 7.1). In fact, ASHRAE 2007 and ADMS compare well at almost all points, except at receptors close to the downwind edge of the building. It may be mentioned that ADMS cannot model the effects of an RTS, and, therefore, only results for building with flat roof are presented for comparisons.

7.5.2 Wind Tunnel and ASHRAE Dilution Results

Figure 7.3a shows wind tunnel dilutions on the rooftop of the low building in the presence of a taller upstream building, for hs = 1 m and M = 1, and compares them

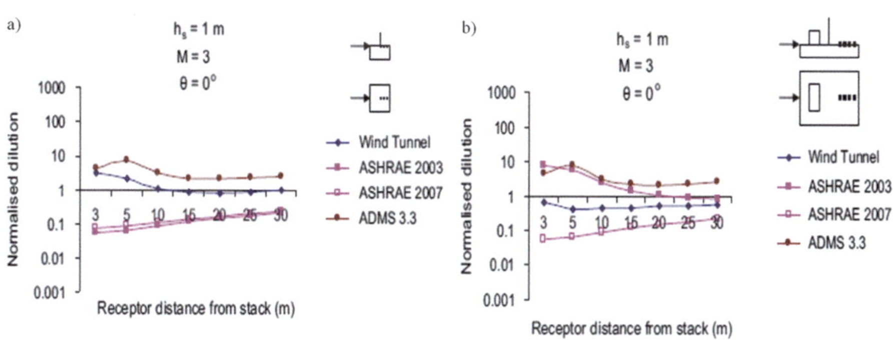

Fig. 7.2 Normalised dilution on rooftop of low building: (**a**) without RTS and (**b**) with RTS (Hajra et al. 2010)

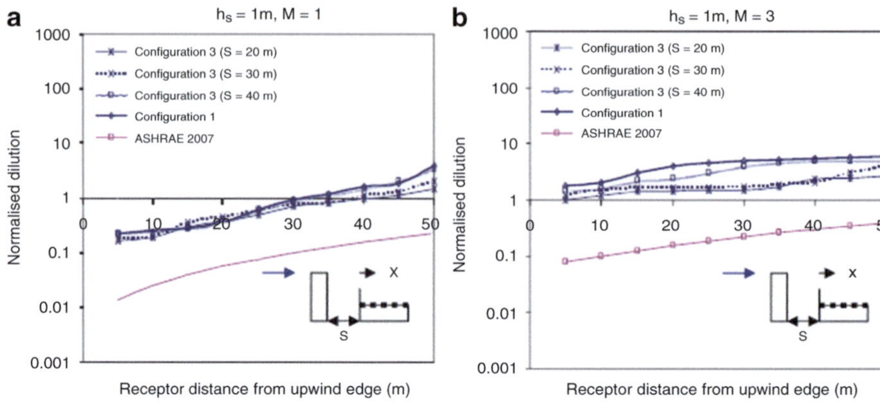

Fig. 7.3 Normalised dilution on rooftop of low building for X = 0: (**a**) M = 1; (**b**) M = 3 (Hajra et al. 2011)

to ASHRAE 2007. Configuration 3 consists of a taller upstream building, while Configuration 1 is an isolated low building. The spacing between buildings was varied from 20 to 40 m. The low-emitting building was 15 m high, and the upstream building was 30 m high. Both buildings were of 50 m across wind dimension. Additional experimental details can be obtained from Hajra et al. (2011). Results show that at M = 1 the wind tunnel dilutions are comparable at all receptors for different spacing. However, at M = 3 the dilutions are comparable only at S = 20 m and S = 30 m (Fig. 7.3b). As the distance increases further to S = 40 m, the dilutions obtained from Configuration 3 become comparable to the isolated case. This is because the plume is no longer engulfed within the recirculation length of the upstream building. ASHRAE (2007) predicts lower dilutions for both configurations and can be applied only for the isolated case. Additionally, the formulations do not incorporate the effects of spacing between buildings.

Figure 7.4 presents comparisons for wind tunnel data from Gupta et al. (2012), Wilson et al. (1998), Schulman and Scire (1991) and ASHRAE (2003, 2007, 2011) models. Wind tunnel data from Gupta et al. (2012) correspond to a low-rise building (15 m high) in an urban terrain, while results from Wilson et al. (1998) and Schulman and Scire (1991) correspond to results for a suburban terrain for a low-rise building of 12 and 15 m, respectively. Generally, the experimental findings from Gupta et al. (2012), Wilson et al. (1998) and ASHRAE (2003) compare well at receptors 20 m beyond the stack, while the data from Schulman and Scire (1991) are somewhat higher (about a factor of 5) than the results from Gupta et al. (2012) due to the difference in experimental conditions used in the two studies. However, ASHRAE 2007 and 2011 predict ten times lower dilutions than wind tunnel data from Gupta et al. (2012), indicating the unsuitability of both models.

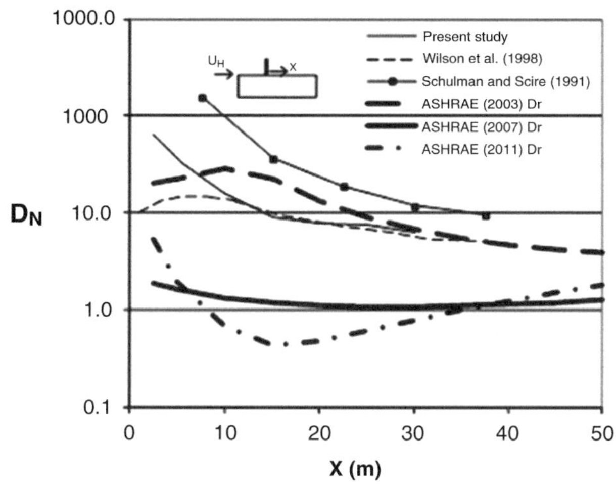

Fig. 7.4 Model validation with concentration data from previous studies for the low-rise building with no rooftop structure for $h_s = 3$ m, and $\theta = 0°$ (Gupta et al. 2012)

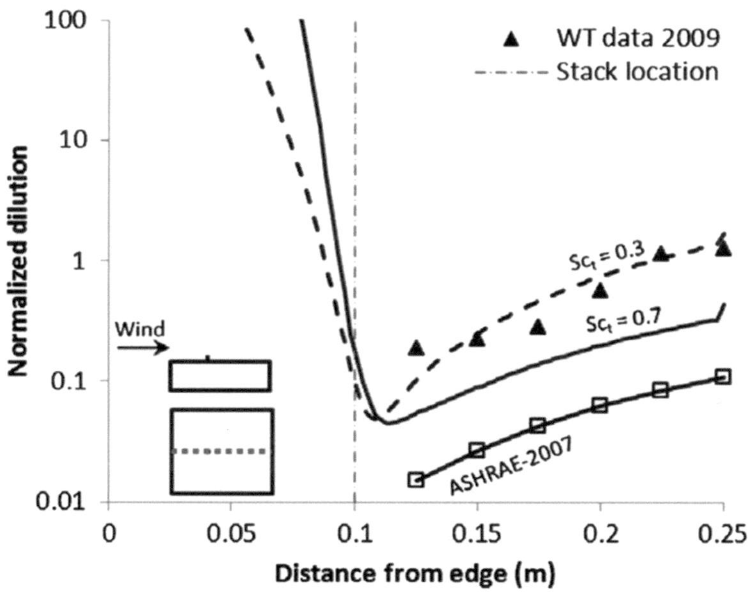

Fig. 7.5 Normalised dilution on isolated building roof for different Sc_t (Chavez et al. 2012)

7.5.3 CFD, Wind Tunnel and ASHRAE Comparisons

Figure 7.5 shows comparisons for wind tunnel data, CFD for Sct values of 0.3 and 0.7 (realisable k-ε model) and ASHRAE 2007 in terms of normalised dilutions for an isolated building with stack height 1 m and M = 1. The building was tested in the boundary layer wind tunnel at Concordia University for an urban terrain (experimental details can be found from Hajra et al. (2010)). Sct is essentially the ratio of the kinematic viscosity to mass diffusivity. Results show that for the centrally located stack, wind tunnel data and CFD compare well for Sct = 0.3 at receptors beyond the stack. However, CFD results are lower than experiment by a factor of 5 for Sct = 0.7 at all receptors, indicating the sensitive nature of the CFD model to changes in Sct. ASHRAE 2007 predictions are overly conservative since the values are lower than wind tunnel data by about 10 times. Interestingly, CFD also predicts dilutions upwind of the stack unlike wind tunnel data, possibly because in the latter case, the instrument cannot detect concentrations of such low value (high dilutions) upwind of the stack. The general behaviour of CFD and ASHRAE suggests that both models must be investigated in detail using experimental data for realistic urban layouts, for future improvements.

7.6 Design Guidelines

The following provides a summary of various design guidelines formulated on the basis of results obtained from some past and very recent studies:

7.6.1 Design Guidelines Based on Field Studies

Field studies were conducted on two of the buildings at Concordia University campus. Tracer gas was released from a low-emitting building, in the presence of a taller upstream building. It was found that the plume was engulfed within the recirculation region of the upstream building, causing the plume to travel towards the leeward wall of the upstream building. This increased the concentrations on the rooftop of the low-emitting building, besides affecting the leeward wall of the taller upstream building. This is one of the most important field studies in the area of near-field pollutant dispersion, since most previous studies were either performed in the wind tunnel for isolated buildings or were focussed on urban air quality modelling for far-field problems. Additional experimental details can be found in Stathopoulos et al. (2004). This subsection describes some of the design guidelines that can be employed for a taller upstream building.

7.6.1.1 Stack Location

For open fetch situations, it is better to place the stack near the centre of the roof. In this way, the leading edge recirculation zone is avoided, thus maximising plume rise. In addition, the required plume height to avoid contact with leeward wall receptors is minimised. For the case of a taller building upwind of the emitting building, the centre of the roof may not be the optimum stack location for receptors on the emitting building. Concentrations over most of the roof can be reduced by placing the stack near the leading edge. However, this stack location will result in higher concentrations on the leeward wall of the adjacent building. Naturally, this depends on the distance between the two buildings.

7.6.1.2 Stack Height

For open fetch situations, increasing the stack height from 1 to 3 m reduces concentrations near the stack by approximately a factor of 2. Far from the stack ($x > 20$ m), the effect is negligible. A stack height of at least 5 m is required to provide significant reduction of concentration at such distances. For an upwind adjacent building, small changes in stack height have little effect on concentration.

7.6.1.3 Stack Exhaust Speed

Increasing stack exhaust speed by a factor of 2.5 reduces concentrations near the stack by the same factor. For distant receptors ($x > 20$ m), the effect of exhaust speed depends on the M value (the ratio of exhaust speed to wind speed). In the low M range ($1.5 < \mathrm{M} < 4.5$), which is typical of wind speeds exceeding 5 m/s, increasing exhaust speed may not be beneficial for distant receptors because the plume rise may not be sufficient to avoid them. On the other hand, for light wind conditions, doubling the exhaust speed may cause M to be high enough so that concentrations are reduced over the entire roof.

7.6.2 Design Guidelines Based on More Recent Studies

More recent studies on the effects of taller upstream configurations and taller downstream configurations have certainly shed more light on the problem of pollutant re-entry into buildings (Hajra et al. 2011; Hajra and Stathopoulos 2012). These studies were performed in the boundary layer wind tunnel at Concordia University, Canada, for two separate cases – upstream building configurations and downstream building configurations. In this context, the building dimensions, flow parameters, wind direction and speed, stack location and height and spacing between buildings were varied. Therefore, this was a more comprehensive study as opposed to the field study at Concordia and limited studies by Wilson et al. (1998). For taller upstream configurations, with wind approaching from the direction of the taller building:

(a) Intakes on emitting building should be placed on its leeward wall if possible.
(b) Intakes should not be placed on leeward wall of upwind building.

According to Hajra et al. (2011), 'When a stack is placed at the upwind edge of the emitting building that lies within the recirculation zone of the upstream

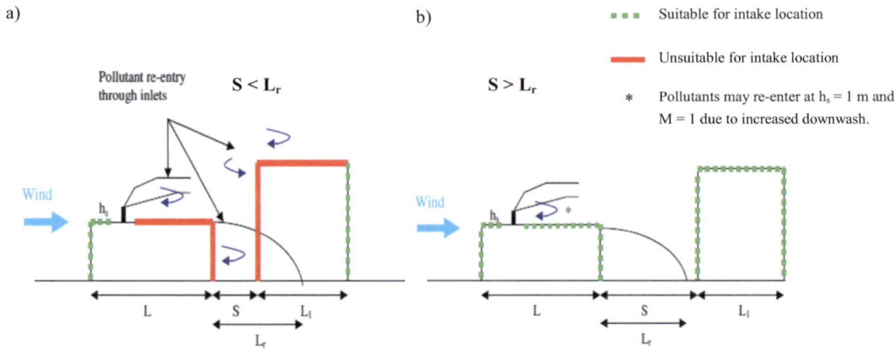

Fig. 7.6 Schematic representation for suitability of intake location at various building surfaces for (**a**) $S < L_r$ and (**b**) $S > L_r$ (Hajra and Stathopoulos 2012)

building, intakes should not be located close to the stack; although they may be placed closer to the leeward wall of the emitting building. For such cases, high stacks and high M values should be used to avoid stack downwash effects'. In general, when a low-emitting building lies within the recirculation zone of a taller upstream building, intakes must be avoided upwind of centrally located stacks, since these locations have high pollutant concentrations due to the geometry of the plume.

Similar guidelines for the safe placement of stack and intakes are also provided for downstream building configurations, as shown in Fig. 7.6 (Hajra and Stathopoulos 2012). Results show that when the downstream building lies within the recirculation zone of the emitting building, intakes should be avoided on building surfaces downwind of the stack (Fig. 7.6a). However, when the downstream building lies beyond the recirculation region of the emitting building, intakes may be safely placed on any building surface (Fig. 7.6b).

7.7 Conclusions

The following conclusions may be drawn from this chapter:

- Dispersion models such as ADMS are more suited for far-field dispersion problems because they cannot model the turbulence caused by buildings and RTS, which greatly influence near-field pollutant dispersion.
- ASHRAE 2003 and 2007 versions were found to be overly conservative for most isolated building cases, while the 2011 version was found to be suitable only for some cases ($M < 3$). In general, all ASHRAE versions are not capable of simulating the effects of adjacent buildings, local topography and rooftop structures.

- CFD cannot simulate the turbulence caused by the building, and, hence, further investigations are required to improve this methodology. In particular, the results are very sensitive to Sct values. Generally, lower values of Sct resulted in comparable dilutions with wind tunnel data for an isolated building.
- Design guidelines from Hajra et al. (2011) and Hajra and Stathopoulos (2012) for the safe placement of stack and intake for upstream and downstream building configurations are a significant contribution in the area of near-field pollutant dispersion studies.

The results obtained are encouraging because they demonstrate the general adequacy of the wind tunnel data to represent real design situations and the limitations of the ASHRAE and ADMS models to predict real dilutions for particular building configurations and stack locations. Dispersion models like ADMS can be improved further, if the formulations used by these models incorporate the effects of wake recirculation region of a building, besides turbulence caused by an RTS and neighbouring buildings. The design guidelines provided in this paper will be helpful to building physicists and practising engineers to tackle a multifaceted complicated problem, for which codes and standards are either mute or extremely general to apply to particular real conditions. It is understandable that additional wind tunnel studies representing a more realistic urban scenario must be conducted to improve future versions of ASHRAE model.

Acknowledgements The financial contribution of the *Institut de recherche Robert-Sauvé en santé et en sécurité du travail* (IRSST), Montreal, Canada, is gratefully acknowledged.

References

ASHRAE (1997) Airflow around buildings. In: ASHRAE handbook fundamentals, Chapter 15. American Society of Heating, Refrigeration and Air-Conditioning Engineers Inc, Atlanta

ASHRAE (1999) Building air intake and exhaust design. In: ASHRAE applications handbook, Chapter 43. American Society of Heating, Refrigeration and Air-Conditioning Engineers, Inc, Atlanta

ASHRAE (2003) Building air intake and exhaust design. In: ASHRAE applications handbook, Chapter 44. American Society of Heating, Refrigeration and Air-Conditioning Engineers, Inc, Atlanta

ASHRAE (2007) Building air intake and exhaust design. In: ASHRAE applications handbook,-Chapter 44. American Society of Heating, Refrigeration and Air-Conditioning Engineers, Inc, Atlanta

ASHRAE (2011) Building air intake and exhaust design. In: ASHRAE applications handbook, Chapter 45. American Society of Heating, Refrigerating and Air-Conditioning Engineering Inc, Atlanta

Blocken B, Stathopoulos T, Carmeliet J, Hensen JLM (2011) Application of computational fluid dynamics in building performance simulation for the outdoor environment: an overview. Build Perform Simul 4(2):157–184

Briggs J (1984) Plume rise and buoyancy effects. In: Randerson D (ed) Atmospheric science and power production, D.O.E./TIC-27601 (DE 84005177). U.S. Department of Energy, Washington, DC

Chavez M, Hajra B, Stathopoulos T, Bahloul A (2012) Assessment of near-field pollutant dispersion: effect of upstream buildings. Wind Eng Ind Aerodyn 104:509–515

Chui EH, Wilson DJ (1988) Effects of varying wind direction on exhaust gas dilution. Wind Eng Ind Aerodyn 31:87–104

Cimoreli AG, Perry SG, Venkatram A, Weil JC, Paine RJ, Wilson DJ (2005) AERMOD: a dispersion model for industrial source applications. Part I: general model formulation and boundary layer characterisation. Appl Meteorol 44:682–693

Gupta A, Stathopoulos T, Saathoff PJ (2012) Evaluation of dilutions models to estimate dilution from rooftop exhausts. ASHRAE Trans 118(Part 1):1–19

Hajra B, Stathopoulos T (2012) A wind tunnel study of the effect of downstream buildings on near-field pollutant dispersion. Build Environ 52:19–31

Hajra B, Stathopoulos T, Bahloul A (2010) Assessment of pollutant dispersion from rooftop stacks: ASHRAE, ADMS and wind tunnel simulation. Build Environ 45:2768–2777

Hajra B, Stathopoulos T, Bahloul A (2011) The effect of upstream buildings on near-field pollutant dispersion in the built environment. Atmos Environ 45:4930–4940

Hajra B, Stathopoulos T, Bahloul A (2013) Performance of ASHRAE models in assessing pollutant dispersion from rooftop emissions. HVAC Res J 20:72–79

Halitsky J (1963) Gas diffusion near buildings. ASHRAE Trans 69:464–484

Higson H, Griffiths RF, Jones CD, Hall DJ (1994) Concentration measurements around an isolated building: a comparison between wind tunnel and field data. Atmos Environ 28(11):1827–1836

Hunt JCR, Robins AG (1982) A model for assessing dispersion of plumes from sources in the vicinity of cuboid shaped buildings. In: Proceedings of the EUROMECH conference 162. New University of Lisbon, Lisbon, Portugal pp 110–116

Petersen RL, Carter JJ, Le-Compte J (2004) Exhaust contamination of hidden versus visible air intakes. ASHRAE Trans 110:600–610

Riddle A, Carruthers D, Sharpe A, Mc Hugh C, Stocker J (2004) Comparisons between FLUENT and ADMS for atmospheric dispersion modelling. Atmos Environ 38:1029–1038

Robins A, McHugh C (2001) Development and evaluation of the ADMS building effects module. Int J Environ Pollut 16:161–174

Saathoff PJ, Stathopoulos T, Wu H (1998) The influence of free-stream turbulence on near-field dilution of exhaust from building vents. Wind Eng Ind Aerodyn 77:741–752

Saathoff P, Gupta A, Stathopoulos T, Lazure L (2009) Contamination of fresh air intakes due to downwash from a rooftop structure. Air Waste Manag Assoc 59:343–353

Schulman L, Scire J (1991) The effect of stack height, exhaust speed, and wind direction on concentrations from a rooftop stacks. ASHRAE Trans 97(Part 2):573–582

Stathopoulos T (1997) Computational wind engineering: past achievements and future challenges. Wind Eng Ind Aerodyn 67&68:509–532

Stathopoulos T, Lazure L, Saathoff PJ (1999) Tracer gas investigation of re-ingestion of building exhaust in an urban environment, IRSST research report no. R-213. Institut de recherche Robert-Sauvé en santé et en securité du travail, Montreal

Stathopoulos T, Lazure L, Saathoff PJ, Gupta A (2004) The effect of stack height, stack location and rooftop structures on air intake contamination: a laboratory and full-scale study, IRSST research report no. R-392. Institut de recherche Robert-Sauvé en santé et en securité du travail, Montreal

Stathopoulos T, Hajra B, Bahloul A (2008) Analytical evaluation of dispersion of exhaust from rooftop stacks on buildings, IRSST research report no. R-576. Institut de recherche Robert-Sauvé en santé et en securité du travail, Montreal

Tominaga Y, Stathopoulos T (2007) Turbulent Schmidt numbers for CFD analysis with various types of flow-field. Atmos Environ 41:8091–8099

Tominaga Y, Stathopoulos T (2011) CFD modelling of pollution dispersion in a street canyon: comparison between LES and RANS. Wind Eng Ind Aerodyn 99:340–348

Wilson DJ (1979) Flow patterns over flat-roofed buildings and application to exhaust stack design. ASHRAE Trans 85(Part 2):284–295

Wilson DJ (1982) Critical wind speeds for maximum exhaust gas re-entry from flush vents at roof level intakes. ASHRAE Trans 88(1):503–513

Wilson DJ, Chui EH (1985) Influence of exhaust velocity and wind incidence angle on dilution from roof vents. ASHRAE Trans 91(Part 2B):1693–1706

Wilson DJ, Chui EH (1987) Effect of turbulence from upwind buildings on dilution of exhaust gases. ASHRAE Trans 93(Part 2):2186–2197

Wilson DJ, Lamb B (1994) Dispersion of exhaust gases from roof level stacks and vents on a laboratory building. Atmos Environ 28:3099–3111

Wilson DJ, Fabris I, Ackerman MY (1998) Measuring adjacent effects on laboratory exhaust stack design. ASHRAE Trans 88:513–533

Chapter 8
Trends in the Field of Quality Assurance of Urban Flow and Dispersion Models

Frank Harms, Denise Hertwig, Bernd Leitl, and Michael Schatzmann

Abstract Caused by synoptic changes and the diurnal cycle, the atmospheric boundary layer is never steady state. The unsteadiness is especially pronounced within and above the urban canopy layer. In former times, it was not possible to take the natural variability of the urban atmosphere properly into account, due to a lack of both computer power for models of adequate sophistication and sufficiently matured measurement techniques. Instead of this, quasi-steady situations were assumed, despite the fact that in reality they do not exist. The situation has improved now. After a brief description of the numerical tools which are presently available, their potential to simulate urban flow and dispersion episodes is assessed. The importance of validating these tools is stressed, and the question of how to obtain reliable validation data is discussed. Using combinations of field and laboratory data for the validation procedure is recommended. Finally, at the concrete example of puff dispersion within the urban canopy layer, it is demonstrated how such data sets can be generated and actually applied.

Keywords Atmospheric variability • Urban boundary layer • Numerical simulation • Model validation • Field and wind tunnel data

8.1 Introduction

Flow and dispersion within and above the urban canopy layer (UCL) has been subject to numerous experimental investigations but is still not well understood (e.g. Oke 1988: Oikawa and Meng 1995; Roth 2000; Rotach et al. 2005; Hanna et al. 2006; Martin et al. 2008; Klein and Young 2011). Due to the generally complex geometrical structure of urban sites, a variety of time and space scales are involved. The urban geometry is heterogeneous and so are the properties of the urban boundary layer flow.

Above the buildings but still within the roughness layer, the flow is continuously adjusting to the ever-changing surface conditions and never-reaching equilibrium.

F. Harms • D. Hertwig • B. Leitl • M. Schatzmann (✉)
Meteorological Institute, Klima Campus, University of Hamburg, Bundesstrasse 55, Hamburg, Germany
e-mail: michael.schatzmann@zmaw.de

Y. Tamura, R. Yoshie (eds.), *Advanced Environmental Wind Engineering*,
DOI 10.1007/978-4-431-55912-2_8

Consequently, the established laws known from flows with long fetch over homogeneous roughness elements (constant fluxes, logarithmic profiles, etc.) are not applicable here.

The situation is even worse within the canopy layer. Here, the flow is channelled by street canyons which, at least in Europe, have many different orientations with respect to the wind direction. It is trivial to note that measured values heavily depend on where in the streets the probes are located. Since the gradients of flow properties are considerable within the UCL, measurements taken a few metre apart from each other might show largely different results (Repschies et al. 2006).

The lack of spatial representativeness is accompanied by a lack of representativeness with respect to time. Even within more or less regular arrays of obstacles and under steady ambient conditions, flow and dispersion properties measured within the UCL are difficult to interpret. Averages over 10 min or even 30 min are usually not ergodic, i.e. repeating an experiment under identical mean boundary conditions would not lead to the same result. As smoke experiments within the UCL reveal, this is caused by low frequency turbulent variations of the flow which make the plume meander (Schatzmann et al. 2006). Results from single measurements are highly uncertain and usually not representative. The degree of uncertainty can be large and must be known before the data can be used for, e.g. model validation purposes.

8.2 Urban Flow and Dispersion Models

Tools which have the potential to accommodate obstacles in a reasonable way are micro-scale meteorological models of prognostic type. These models are based on the Navier–Stokes equation. To directly solve the equation in a turbulent flow requires a very fine grid to capture all the relevant scales, down to the so-called Kolmogorov scale (usually less than a millimetre). Furthermore, a time-dependent solution over a sufficiently long period (many hours) is needed to yield stable time averages of the flow variables. This approach is called direct numerical simulation (DNS). As its computational demand is too high for the domain sizes and Reynolds numbers typically encountered in atmospheric boundary layers, DNS is not applicable here.

The computational demand can be substantially reduced when the time-dependent equations are solved on a grid that is fine enough (less than a few meter in the case of a city quarter) to resolve the larger atmospheric eddies. This approach is called large-eddy simulation (LES). The small scales are formally removed from the flow variables by spatially filtering the Navier–Stokes equations. The influence of the small scales then appears as sub-filter stresses in the momentum equation. Since the large eddies are always unsteady, LES models require input conditions which are time dependent as well. Whilst being less demanding than DNS, LES still requires significant computer resources which go beyond the capabilities many users presently have.

In view of this, the still most widespread method used for the computation of turbulent atmospheric flows is the Reynolds-averaged Navier–Stokes (RANS) approach. Within this approach, the equations are averaged in time over all turbulent scales, to directly yield the statistically steady solution of the mean and turbulent flow variables. Like LES, the averaging leads to additional terms in the momentum equation known as the Reynolds stresses. They represent the effects of the turbulent fluctuations on the averaged flow and have to be parameterised. This is the task of turbulence closure models. In most models which are presently in use, the Reynolds stresses are assumed to depend linearly on the strain rate, as do the molecular stresses. The eddy viscosity appears as a proportionality factor that can be calculated using additional differential equations for the various order moments. Many modellers regard the two-equation turbulence closure schemes which solve differential equations for the turbulent kinetic energy, k, and the dissipation rate, ε, a good compromise between universal validity and operating expense. In particular, the standard $k - \varepsilon$ model is widely used in engineering and micrometeorological applications, despite the fact that it produces systematic errors as, e.g. too much turbulent kinetic energy in regions of stagnant flow, etc. Several modifications have been proposed which ease this problem but most times on the expense of the quality of other flow property predictions.

A further option is the use of standard turbulence models from the RANS approach in time-dependent simulations. Contrary to LES, the averages are defined as ensemble or as time averages over small time intervals, although the latter definition leads to more additional terms in the momentum equation than in the case of ensemble averaging. This approach is also known as unsteady RANS (URANS). It is used in standard meteorological meso- and macro-scale applications (weather forecasts, etc.). URANS models are driven by time-dependent boundary conditions. They account, e.g. for different land uses with different radiation budgets and provide time-dependent predictions. In the context of urban applications, URANS models are under development. However, much more powerful computers than presently available and substantial research efforts will still be needed before the first reliable unsteady obstacle resolving URANS predictions for urban-scale dispersion problems will become available.

Another class of models which are still in use are diagnostic flow models. They do not use the Navier–Stokes equation but are based solely on mass conservation. The important influence of pressure gradients and forces on the flow development can only indirectly (empirically) be taken into account. These models start with a first guess of the three-dimensional flow field that is subsequently modified until the divergence of the flow falls below a chosen limit. For a given obstacle array, many different mass-conserving flow fields can be found, depending on the particular choice of the initial flow field and the ‘tuning of knobs’ inside the model which, e.g. determine whether the fluid at a specific position moves over or goes around an obstacle. Diagnostic models may be helpful in analysing known cases, in particular when a good set of observed velocity data is available for the area of interest. However, they cannot themselves be regarded as tools for the prediction of new cases.

To determine the dispersion of pollutants, diagnostic models are usually combined with Lagrangian models. These are pure dispersion models, i.e. they follow individual plume particles and model their paths on the basis of a random walk process. The number of particles which arrive at a given grid cell is an indicator for the concentration in that cell. Lagrangian models need a complete mean and turbulent flow field as model input, which is usually delivered in form of 3D-gridded fields by either a diagnostic or prognostic model. Consequently, the quality of Lagrangian predictions is closely linked to the quality of the turbulent velocity field provided by the flow model.

Gaussian models, finally, are mentioned here only for the sake of completeness. These models are pure dispersion models and applicable only for pollutant emissions into uniform and steady-state atmospheric flows (e.g. tall stack releases in flat, unobstructed terrain and averaged over a large number of atmospheric conditions). Although there are some Gaussian models with modifications aimed to accommodate vertically changing mean boundary layer properties or one or a few buildings, it is generally accepted that these models are not appropriate for predicting flow and concentration in complex structured urban or industrial areas.

Which model is fit for the purpose depends, of course, on the particular application. For the most common case, the prediction of flow and dispersion in urban environments for a given scenario, one can be more specific. In view of what was explained before, it occurs that only CFD models of RANS or LES type can be regarded as predictive tools. The other model types are either not yet capable of doing the job (DNS, URANS) or lack the potential for proper work in flow fields disturbed by industrial or urban obstacle arrays (Gaussian and diagnostic models).

8.3 Urban Flow and Dispersion Models

When a model has the potential to simulate flow and dispersion episodes in complex terrain, it does not mean that it is actually able to do this in a proper way. Whether the model output is in agreement with observation has to be proven in model validation exercises. Validating models is an art in itself (Oberkampf et al. 2004; Roache 1998). With respect to urban models, the most pertinent guidance has probably be given by the network QNET-CFD (Castro 2003), by the Architectural Institute of Japan (Yoshie et al. 2007; Tominaga et al. 2008), by the European action COST 732 (Britter and Schatzmann 2007a, b; Franke et al. 2007; Schatzmann et al. 2010) and very recently with focus on LES models by Hertwig (2013) and the European COST action ES 1006 (see http://www.elizas.eu). The backbone of any validation work is the existence of data which have sufficient quality that they can be regarded as a standard.

COST 732 stresses several constraints which are typically faced during normal CFD applications and in particular during the validation process for this model type. With respect to RANS, the most important constraint is that steady-state models require validation data from experiments which were taken under steady-state atmospheric conditions. However, the atmosphere is intrinsically time dependent and never steady state. The weather is continuously changing, due to both the atmospheric circulation and the diurnal cycle. The fluctuations and gusts in the incoming wind interact with equally important vortex shedding from buildings to control the local flow fields and the dispersion of pollutants.

To overcome this problem, quasi-steady situations are defined which are composed of mean values averaged over, e.g. 30 min. However, as was shown in Schatzmann and Leitl (2011) at the example of a detailed analysis of data from an urban monitoring station for a full year, the time scale of the naturally occurring turbulent structures significantly exceeds such short averaging periods. As a consequence, the commonly determined short-time mean values measured inside the urban canopy layer have the character of random samples only (or snapshots, to put it more bluntly). Depending on the wind direction, the variability between seemingly identical cases can be larger than a factor of 2. To simply increase the sampling time would not solve but worsen the problem since over periods longer than 30 min a systematic trend in meteorological conditions due to the diurnal cycle has to be expected. Wind tunnel and LES simulations show that many hours of constant weather conditions would be needed until a time series becomes ergodic and delivers always the same mean value. In view of this, it has to be concluded that episodic field measurements cannot be representative at locations with highly fluctuating flow properties as it is the case in urban environments.

A second problem caused by the naturally occurring atmospheric variability concerns the determination of representative input conditions for the model runs. In most field experiments, only one reference station exists that can be used for characterising the meteorological situation and the model input. Fortunately, however, there are a few exceptions, most prominent amongst them the Joint Urban 2003 experiment in Oklahoma City (Allwine et al. 2004). In this experiment, the wind vector was measured with high temporal resolution at multiple locations in parallel. Two of the instruments with which the wind was measured were mounted onto poles well above the rooftops of buildings located outside the Business District and about 1 km apart from each other. The data from these two sites (PNNL and PWIDS) are often selected to determine the inflow profiles for numerical model runs. At least for southerly and westerly wind directions, the flow approaching the instruments is regarded as being undisturbed.

Hertwig (2008) compared for several periods the wind velocities and directions which were measured simultaneously at the two stations. A typical result is shown in Fig. 8.1. The data are presented as they were measured, i.e. no corrections were made for the different measurement heights (PWIDS 49 m and PNLL 37 m above ground). In order to reduce the high-frequency scatter, the data were averaged over 5 min. As can be seen in the figure, there is a significant variability in the wind velocities, whereas the wind direction is comparatively constant. This is not always the case; other examples show just the opposite or a variation of both. Not very surprising, the curves in Fig. 8.1 differ from each other, and a correction of the data for different measurement heights would not diminish but increase the differences. Although not the raw data but already 5 min mean values are presented, the data do

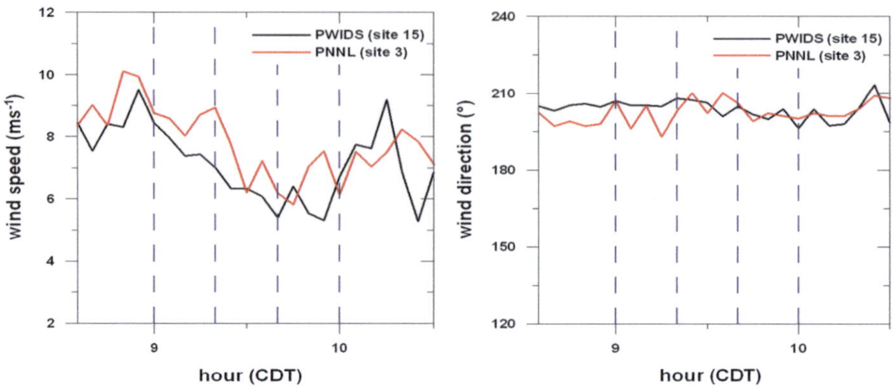

Fig. 8.1 Wind speeds (*left*) and directions (*right*) averaged over 5 min measured at the stations PWIDS 15 and PNNL 3 during the Joint Urban 2003 campaign IOP3. The *dashed lines* mark 20 min intervals (From Hertwig 2008)

not change in synchrony, which indicates that there are large-scale turbulent structures superimposed to the mean flow.

Assumed that 20 min averages would be used to determine the input for a model run, it becomes very obvious that the models would predict quite distinct flow and, above all, concentration fields, depending on from which measurement site the input data were chosen. In the Oklahoma field experiment, more than 100 anemometers, Sodars, etc. were deployed and measured in parallel. The data of many of them would likewise be suitable to be used as reference velocities. All of them deliver different model inputs, and the differences are much larger than the instrument uncertainty. A model user could select the input which leads to the best fit. To regard such a test as a serious proof for model quality would certainly not be justified.

The ambiguity in the right choice of input data for model simulations concerns every field data set, not only those in which the model input can be selected from several meteorological towers. In experiments with one reference station only, it simply remains unknown how representative this tower measurement really is. The large turbulent elements embedded in the mean flow create changes from location to location and from averaging interval to averaging interval which could be determined only if the mean weather remained constant or a sufficiently large ensemble of short-time means could be collected and statistically analysed.

A third difficulty which has been encountered in the search for reliable validation data sets originates from the fact that resources are limited and field measurements usually are made at a few selected points only. A CFD model, on the other hand, requires at the inflow boundary of the domain a complete data array, and this not only for the mean velocity but also for more complex properties like the turbulent kinetic energy or the dissipation rate. Provided the measurement was made with a high-resolution instrument that delivered a velocity time series (which is usually not the case in older field data sets), this information is available for the measurement position only. It needs to be extrapolated laterally and vertically which requires additional assumptions to be employed (e.g. the assessment of a roughness length, a logarithmic wind profile in the surface layer, lateral homogeneity, etc.). In view of what was explained before, all these assumptions are most likely not justified over short averaging intervals but presume the existence of long periods of constant weather, the existence of a boundary layer in equilibrium with the underlying surface, etc. As a consequence, the model is fed with input which most likely does not sufficiently correspond to the flow or concentration measurements which are used in the validation process to check the accuracy of the model output.

8.4 Expected Trends in the Validation of Urban Models

Since it is unfair to blame a model for "flaws" contained in the data and which originate from the inherent variability of the atmosphere, a strategy is needed in order to cope with the situation. A tendency can be recognised to base urban model validation work predominantly on data obtained in boundary layer wind tunnel experiments. At least for some meteorological situations (moderate to high wind speeds, neutral stratification), data which fulfil the pertinent similarity criteria can be generated. Other advantages of wind tunnel experiments are that the complexity of the experimental set-up can be chosen at will and that the physical boundary conditions can be controlled and kept constant over a sufficiently long period of time. Quasi-steady conditions, as they are postulated for most of the model types, can be easily achieved, and statistically representative measurements can be realised even for systematically varying boundary conditions. Complete arrays of data can be produced which correspond to those provided by the numerical model. By means of repetitive measurements under similar boundary conditions and time series analyses, the uncertainty and variability of the data can be specified. Precondition is, of course, that the wind tunnel boundary layer characteristics replicate those of the real atmosphere and that this is proven not only with respect to mean values but with respect to the statistics of the underlying turbulent properties as well (see, e.g. Leitl 2000; or Leitl and Schatzmann 2010).

Despite the benefits laboratory data have, they are the result of another model which has its own restrictions. The data are related to the physical reality, but they do not represent the reality itself. To master this challenge, COST 732 decided to accept only data from laboratory experiments in which a real field situation was replicated. This gives the opportunity to directly compare the wind tunnel results with field data. In addition, such combined data sets can extend the value of field experiments by adding, e.g. information for weather situations not covered during the experimental campaign. Examples for such data sets suitable for the validation of RANS CFD models are meanwhile available and can be downloaded, e.g. from the COST 732 actions homepage under http://www.mi.zmaw.de/index.php?id=484.

8.5 Demonstration of the Method at a Specific Example

With large-eddy simulation models becoming readily available for urban applications (Patnaik et al. 2009), the demand for data sets with high temporal and spatial resolution is intensified. Since LES models produce time series of flow and dispersion properties, they require test data which consist of time series as well, and this not only for the properties of interest but for the boundary conditions as well. These time series need to be analysed, and comparisons must be made for more but just mean values, i.e. for higher-order statistics, correlation information in terms of integral length scales, energy density spectra, etc. A full account of performance quality measures for LES models has been given in Hertwig (2013) and in the publications by the COST action ES 1006 (see http://www.elizas.eu). It seems to be hardly feasible to obtain time series for such detailed investigations in urban field experiments which shifts the focus again to wind tunnel modelling.

The key role of combined data sets in the context of LES model testing is demonstrated by Harms (2010) and Harms et al. (2011) at the example of a real case, the dispersion of puffs as they might be released in case of an accident. Harms replicated parts of the field campaign Joint Urban 2003 which was carried out in Oklahoma City (Allwine et al. 2004) in the scale of 1:300 in the large boundary layer wind tunnel of Hamburg University. A drawing of the facility is shown in Fig. 8.2. The 25 m long wind tunnel provides an 18 m long test section equipped with two turn tables and an adjustable ceiling. The cross section of the tunnel measures 4 m in width and 2.75–3.25 m in height depending on the position of the adjustable ceiling.

Whilst in the test section free stream wind speeds of more than 20 m/s can be realised, the typical wind velocities chosen for the experiments were in the range of 5–15 m/s, thereby achieving Reynolds number independence. The model boundary layer flow was generated by a carefully optimised combination of turbulence generators at the inlet of the test section and roughness elements on the floor, thereby following the procedure outlined in Schatzmann and Leitl (2011). The large cross section of the tunnel allowed the reproduction of atmospheric eddies with sizes of about 1 km which corresponds to turbulent wind fluctuations with time scales of more than 1 h. Figure 8.3 shows the physical model of the Central Business District of Oklahoma City mounted in the wind tunnel. The diameter of the detailed wind tunnel model was 3.5 m, corresponding to 1050 m at full scale.

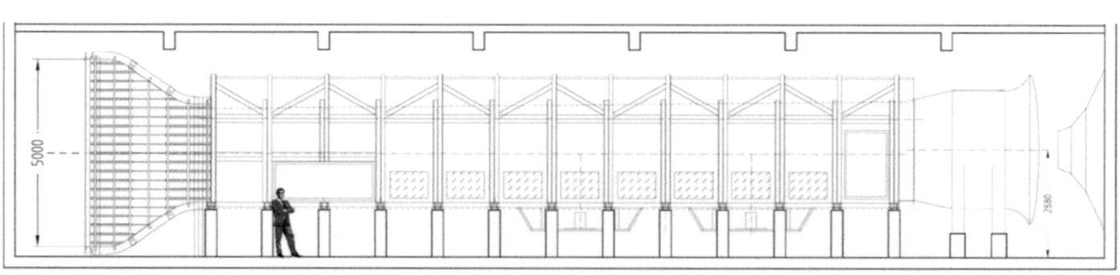

Fig. 8.2 Drawing of the large boundary layer wind tunnel facility 'WOTAN' of the University of Hamburg

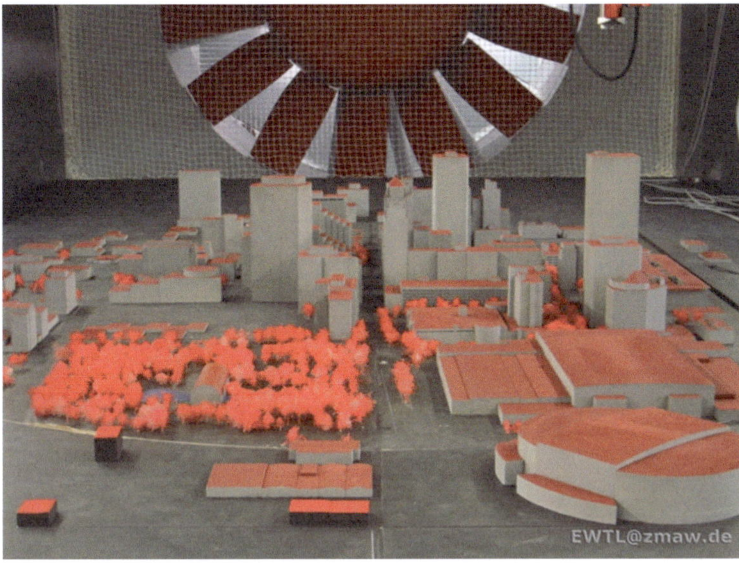

Fig. 8.3 Wind tunnel model of the Central Business District of Oklahoma City

For dispersion modelling and measurements, several point emission sources were flush mounted in the model floor. The circular release area had a diameter of 7 mm (model scale), corresponding to 2.1 m at full scale. In order to avoid the forming of a significant vertical jet at higher emission rates, the source area was covered by a lid, 3.5 mm (1.05 m full scale) above-ground level. The modelled release locations were chosen in accordance with some of the release sites used during the JU 2003 field campaign. In order to simulate instantaneous puff releases, a continuous by-pass flow of tracer gas was temporarily switched to the source by means of a fast solenoid micro-valve. With this setup, the release rate could be kept absolutely constant for repetitive releases lasting much less than a second at model scale. The precise repeatability of releases was verified by extensive systematic test series prior to the actual model tests.

Extensive flow measurements with high temporal and spatial resolution were carried out. By using laser Doppler anemometry (LDA), component-resolved velocity time series at sampling rates of several hundred Hertz were generated at more than 1700 measurement locations. The flow measurements were complemented by fast flame ionisation detector (FFID) measurements, thereby obtaining concentration versus time traces with frequency responses in the range of 100 Hz. Both the LDA and the FFID were mounted to a traverse system. Whereas sampling took place near to the ground, the instruments were positioned well above the urban canopy to avoid the flow to be disturbed.

Figure 8.4 illustrates the variability of individual puff concentration time traces recorded at a measurement location within the urban canopy for seven identical releases from a source located upstream. The instants of releases are indicated by

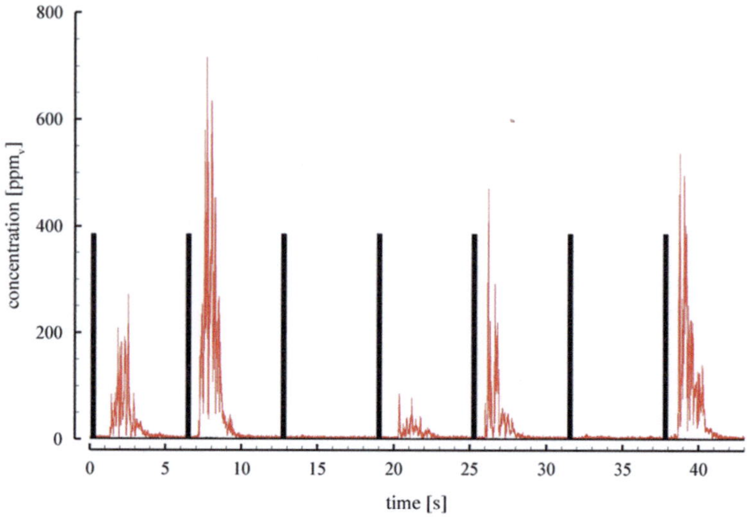

Fig. 8.4 Sequence of measured concentration signals from seven identical puff releases. The release times are indicated by the *black bars*

the black bars, and the red line states the measured concentration in ppm$_v$ at the measurement location. The figure shows that two of the seven released clouds completely miss the probe position, whilst the measured concentrations for the other five releases differ significantly. Since the mean boundary and source conditions were identical for all releases, the observed differences must have been caused by the turbulent flow field. As is evident from Fig. 8.4 large ensembles of puffs need to be collected and analysed in order to characterise the dispersion of puffs properly, individual puffs are random events.

Although a 'mean puff' could be defined from a sufficiently large ensemble of releases, a solely pattern-based comparison of puffs from model simulations and experiments would be too time-consuming. In order to simplify the comparison, a puff can be characterised by a number of parameters such as arrival time (at), peak time (pt), leaving time (lt) and dosage (dos) or peak concentration (pc). Each of the parameters illustrated in Fig. 8.5 can be calculated for each of the individual puff signals recorded at a given measurement location. For detecting the arrival time and leaving time from a recorded time series, a dosage-based method was used. As indicated in Fig. 8.5, the arrival time and leaving time define the time interval after the release when 5 % and 95 % of the total dosage of a puff reached the measurement location, respectively.

Plotting a sufficiently large ensemble of derived puff parameters, a well-defined and sufficiently smooth frequency distribution can be achieved. Figure 8.6 shows a frequency distribution plot generated from wind tunnel measurements for the peak time parameter. In order to estimate the number of repetitive releases required for achieving a reasonable confidence interval of the ensemble-averaged results, a series of pretests was carried out. For a variety of possible measurement locations,

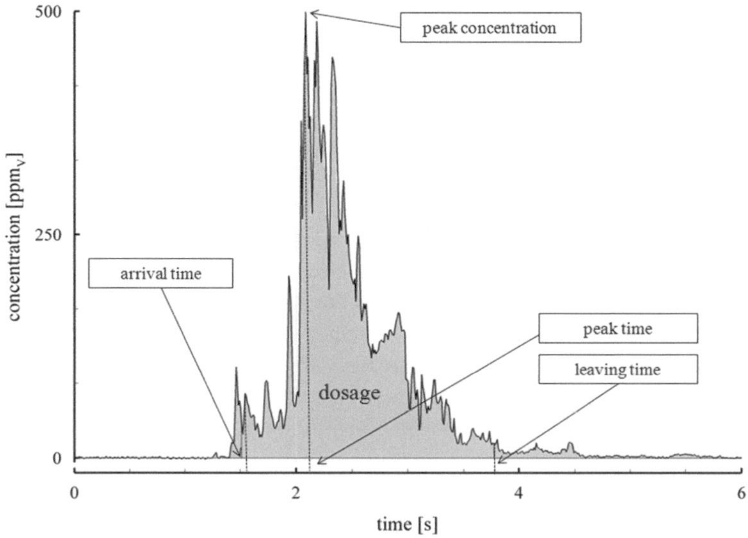

Fig. 8.5 Definition of puff parameters. The arrival time is defined as the time after release when the dosage exceeds the threshold of 5 %. The leaving time is defined as the time when 95 % of the total dosage is reached

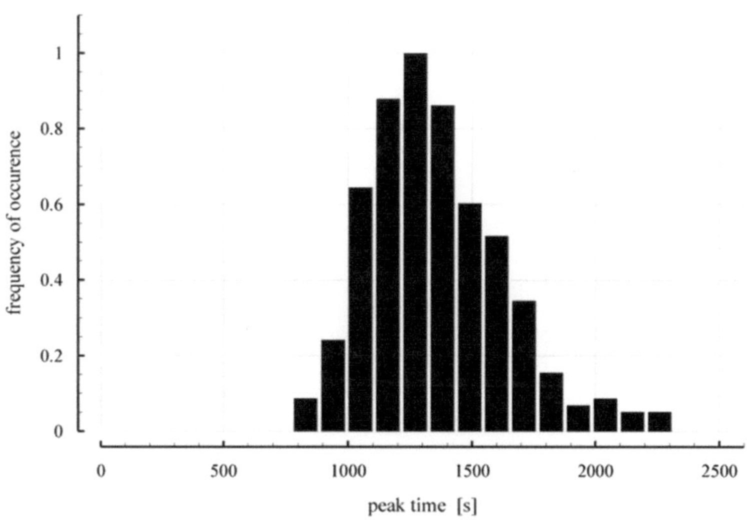

Fig. 8.6 Frequency distribution plot of the 'peak time' parameter measured at a relatively undisturbed street crossing (Robert S Kerr Avenue – Broadway Avenue) based on 200 identical releases from a source located upwind of the city centre

several hundreds of individual releases were carried out, and ensemble-averaged values of puff dispersion parameters were calculated for gradually increasing ensemble sizes. Additionally for each ensemble size, the mean values were calculated. It was found that a minimum of about 200 releases were required in order to reach a confidence interval sufficient for model validation purposes.

Figure 8.7 shows a typical result of a convergence analysis for the measured puff travel time. In this case, 281 identical puffs were released from a source located in the centre of the Park Avenue, modelling a mean wind direction of 180° and a mean wind speed of 2.4 m/s at 80 m height above ground just upstream of the Central Business District. The measurement location was located one street canyon downstream from the source. The Figure illustrates the uncertainty in defining the mean arrival time if the ensemble size is limited. As expected, the uncertainty in defining a repeatable mean value decreases with increasing ensemble size. In this particular case, 200 releases allow defining the mean arrival time with an uncertainty of ±5 %. This uncertainty increases to at least ±15 % if the mean value is calculated from 50 releases only.

For field measurements, it has to be considered that due to changing boundary conditions, the number of puff measurements which can be carried out under similar boundary conditions is typically much less than 50 releases. During the Joint Urban 2003 campaign, at most four puffs were released under similar boundary conditions. For the given dispersion situation, Fig. 8.7 indicates that the mean arrival time obtainable from four releases only would, for this specific source/receptor arrangement, have an uncertainty of ±60 %.

As illustrated by the black line in Fig. 8.7, the reduction of the uncertainty with increasing ensemble size can be described by the relation

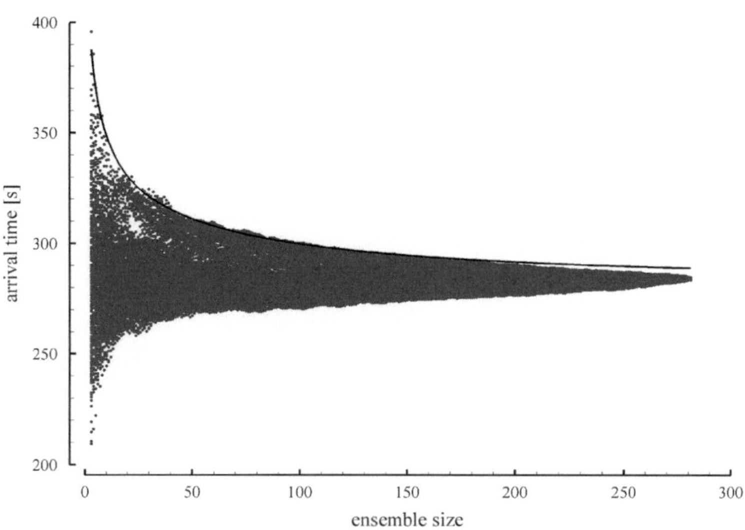

Fig. 8.7 Mean arrival time calculated for different ensemble sizes. The tracer was released from a source located in Park Avenue. The measurement location was located one street canyon further downstream. The mean wind direction was 180°

$$y = A + \frac{B}{\sqrt{n}}$$

in which y describes for each ensemble size n the maximum difference to the mean arrival time which was calculated from all 281 puffs. A and B are curve parameters which values vary depending on the considered dispersion scenario. The finding is in line with statistical theory according to which the reduction of the uncertainty with increasing ensemble size n is proportional to $1/\sqrt{n}$.

The wind tunnel measurements document that the uncertainty in defining mean values depends on other parameters too. For measurements close to the source location, the observed concentration gradients are stronger, and an increased number of releases are required to define the desired mean values with the same statistical confidence. In addition, it was found that at the same measurement location, the uncertainty range differs for the various puff parameters. The dosage and peak concentration seems to be the parameters with the biggest uncertainty levels. This information is essential when mean values from wind tunnel tests, field experiments or numerical results are compared with each other.

Subsequently, the wind tunnel results are compared with corresponding field data and results from the LES model FAST3D (Patnaik et al. 2009). Figure 8.8a shows a frequency distribution plot generated from 200 puff releases measured in the wind tunnel. The puffs were released from a source upstream of the city centre next to the botanical garden (Fig. 8.8b). The measurement location was located in the Robert S Kerr Avenue. The plot shows the 'peak time' parameter in comparison with three corresponding releases from IOP (intensive operation period) 3 of the Joint Urban 2003 field campaign. The mean wind speeds during the three field releases varied. In order to compare the travel time of the field puffs with each other and with the wind tunnel data, all results were scaled to the same mean wind speed

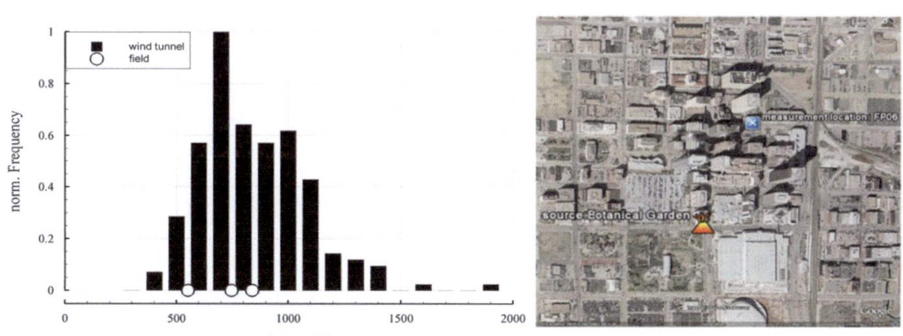

Fig. 8.8 (**a**) Frequency distribution plot generated by 200 wind tunnel measurements for the 'peak time' parameter compared to three releases from IOP 3 of the Joint Urban 2003 field campaign. (**b**) The puffs were released from the botanical garden source. The measurement location was in the Robert S Kerr Avenue (Picture from Google Earth)

of 2.5 m/s expected to be present at 80 m height above ground just upstream of the Central Business District.

The comparison indicates that the peak times calculated for the three field puffs are entirely covered by the wind tunnel distribution of possible peak times. The plot demonstrates furthermore that the complete bandwidth of possible results for the given situation cannot be estimated reliably based on a few field releases only.

Altogether, 30 different dispersion scenarios of the Joint Urban 2003 campaign were analysed in the wind tunnel. A dispersion scenario is given by the combination of the source location, the measurement location and the mean wind speed. For each dispersion scenario, the results of wind tunnel and field measurements were compared. In each scenario and for all puff parameters, the field results were covered by the wind tunnel distributions.

Figure 8.9a shows a comparison of the 'peak time' parameter derived from 200 wind tunnel releases and 60 puffs released in the LES-based numerical model fast 3D. The tracer was released again from the source location 'botanical garden', and measurements took place at FP05 (Fig. 8.9b), a street crossing within the most densely built-up area. Wind direction was adjusted to 180° with a mean wind speed of 2.5 m/s measured at 80 m above ground in the undisturbed approach flow. Comparing the mean values revealed a fair agreement. The difference of the mean peak times is less than 6 %. In addition, Fig. 8.9a illustrates for this case an almost perfect matching for the frequency distributions of wind tunnel data and the results of the numerical simulation. This qualitative examination is confirmed by the results of the Mann–Whitney test, which determines that the difference between the medians is not statistically significant. In this case, there were a sufficiently large number of trials (puffs) in the numerical simulations which means that we can have a high degree of confidence that both distributions are samples from the same population.

Figure 8.10a shows a second comparison for the ensemble-averaged 'peak time' parameter calculated from wind tunnel measurements and corresponding numerical

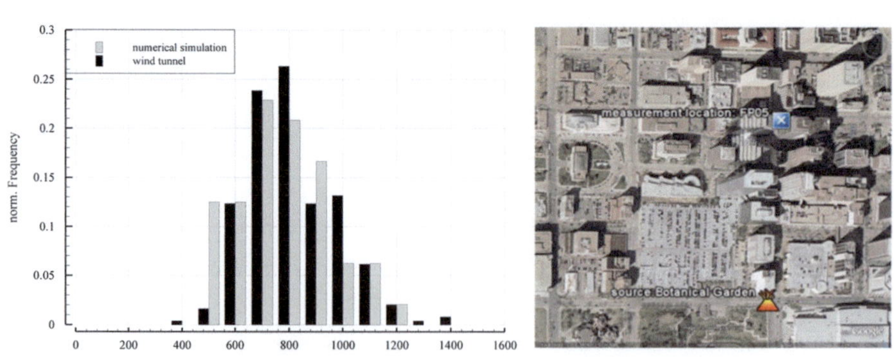

Fig. 8.9 (**a**) Frequency distribution plot for the 'peak time' parameter generated by 200 wind tunnel and 60 numerical releases. (**b**) The tracer was released at the botanical garden source and measured at FP05 (Picture from Google Earth)

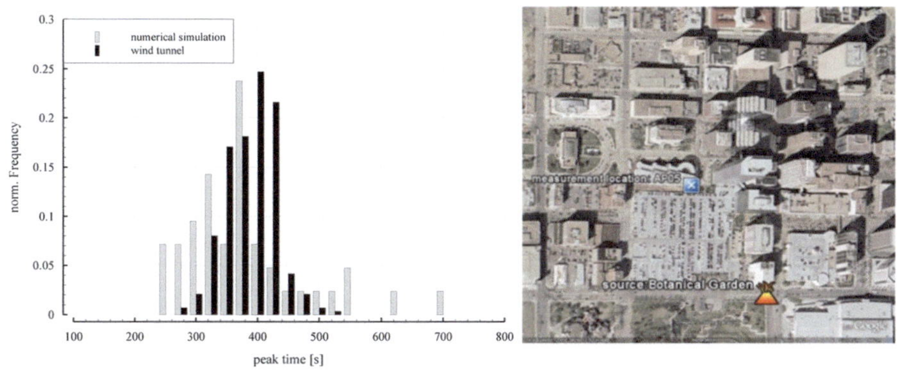

Fig. 8.10 (**a**) Frequency distribution plot for the 'peak time' parameter generated by 200 wind tunnel and 60 numerical releases. (**b**) The tracer was released at the botanical garden source and measured at AP05 (Picture from Google Earth)

simulations. The approach flow conditions and the source location are kept the same as in the previous example. However, the measurement location is now located at the downwind edge of a large parking lot (Fig. 8.10b). The mean peak time of 375 s calculated from 200 wind tunnel releases is very similar to the value of 367 s calculated from the 60 releases simulated in the numerical model. The difference of the mean values is just 2 %. However, the comparison of the corresponding frequency distributions (Fig. 8.10a) illustrates differences between wind tunnel and numerical results. For this case, the result of the Mann–Whitney test indicates that the difference between the medians of the numerical wind tunnel distributions of the peak times is considered extremely significant; and therefore, the numerical simulations do not agree for this metric at this source measurement location pair. Other metrics (e.g. arrival time) also show disagreement here.

Examination of plumes generated by the numerical simulations shows that the measurement location is right at the edge of the plume envelope and fewer puffs reach here than in the 'good' agreement case, which is down the centreline of the plume. In general however, the agreement between the LES model predictions and the wind tunnel results over all the cases was found to be fair. These results clearly indicate the danger of selecting a single plume characteristic (e.g. mean peak time) to evaluate the quality of numerical results for validation purposes.

8.6 Conclusions

The paper deals with the fact that the atmosphere is never steady state and which implications that has with respect to urban flow and dispersion modelling. The numerical model types presently available are briefly introduced, and it is stated that the trend is towards CFD models. Above all, LES-CFDs have the potential to cope with flows heavily disturbed by buildings and other obstacles. However, since a potential is not a proven ability, the models need to be properly validated. The difficulty of finding reliable validation data is emphasised, and it is recommended to base the validation work on data from combinations of field and wind tunnel experiments. The usefulness of this recommendation is finally demonstrated at the example of puff dispersion within an urban canopy layer.

Acknowledgements The experimental data was gathered within a DTRA-funded joint research project between the Naval Research Laboratory in Washington DC and the Environmental Wind Tunnel Laboratory at the University of Hamburg.

References

Allwine KJ, Leach MJ, Stockham LW, Shinn JS, Hosker RP, Bowers JF, Pace JC (2004) Overview of Joint Urban 2003 – an atmospheric dispersion study in Oklahoma City. Symposium on planning, nowcasting, and forecasting in the Urban zone, Seattle, 11–15 Jan 2004. Am Meteor Soc J 7.1

Britter R, Schatzmann M (eds) (2007a) Background and justification document to support the model evaluation guidance and protocol document. COST Office, Brussels. ISBN 3-00-018312-4

Britter R, Schatzmann M (eds) (2007b) Model evaluation guidance and protocol document. COST Office, Brussels. ISBN 3-00-018312-4

Castro IP (2003) CFD for external aerodynamics in the built environment. QNET-CFD Netw Newsl 2(2):4–7

Franke J, Hellsten A, Schlünzen H, Carissimo B (eds) (2007) Best practice guideline for the CFD simulation of flows in the urban environment. COST Office, Brussels. ISBN 3-00-018312-4

Hanna SR, Brown MJ, Camelli FE, Chan ST, Coirier WJ, Kim S, Hansen OR, Huber AH, Reynolds RM (2006) Detailed simulations of atmospheric flow and dispersion in Downtown Manhattan: an application of five computational fluid dynamics models. Bull Am Meteorol Soc 87(12):1713–1726

Harms F (2010) Systematische Windkanaluntersuchungen zur Charakterisierung instationärer Ausbreitungsprozesse einzelner Gaswolken in urbanen Rauigkeitsstrukturen. PhD thesis, University of Hamburg (in German)

Harms F, Leitl B, Schatzmann M, Patnaik G (2011) Validating LES-based flow and dispersion models. J Wind Eng Ind Aerodyn 99:289–295

Hertwig D (2008) Dispersion in an urban environment with a focus on puff releases. Study Project, Meteorological Institute, University of Hamburg

Hertwig D (2013) On aspects of large-eddy simulation validation for near surface atmospheric flows. PhD thesis, University of Hamburg

Klein PM, Young DT (2011) Concentration fluctuations in a downtown urban area. Part I: analysis of Joint Urban 2003 full-scale fast-response measurements. Environ Fluid Mech 11:23–42

Leitl B (2000) Validation data for microscale dispersion modelling. EUROTRAC Newsl 22:28–32

Leitl B, Schatzmann M (2010) Validation data for urban flow and dispersion models: are wind tunnel data qualified? The fifth international symposium on computational wind engineering (CWE2010), Chapel Hill, 23–27 May 2010

Martin D, Price CS, White IR, Nickless G, Dobre A, Shallcross DE (2008) A study of pollutant concentration variability in an urban street under low wind speeds. Atmos Sci Lett 9:147–152

Oberkampf WL, Trucano TG, Hirsch C (2004) Verification, validation, and predictive capability in computational engineering and physics. Appl Mech Rev 57:345–384

Oikawa S, Meng Y (1995) Turbulence characteristics and organized motion in a suburban roughness sublayer. Bound-Lay Met 74:289–312

Oke TR (1988) The urban energy balance. Prog Phys Geogr 12:471–508

Patnaik G, Boris J, Lee MY, Young T, Leitl B, Harms F, Schatzmann M (2009) Validation of an LES urban aerodynamics model with model and application specific wind tunnel data. In: Proceeding of the 7th Asia-Pacific conference on wind engineering, Taipei, 8–12 Nov 2009

Repschies D, Leitl B, Schatzmann M (2006) Validation of urban dispersion models: how representative are locally measured wind velocities? In: Proceedings of the 10th annual George Mason University Conference on atmospheric transport and dispersion modeling. Fairfax, 1–3 Aug 2006

Roache PJ (1998) Verification and validation in computational science and engineering. Hermosa Publishers, New Mexico

Rotach M, Vogt R, Bernhofer C, Batchvarova E, Christen A, Clappier A, Feddersen B, Gryning SE, Martucci G, Mayer H, Mitev V, Oke TR, Parlow E, Richner H, Roth M, Roulet YA, Ruffieux D, Salmond JA, Schatzmann M, Voogt JA (2005) BUBBLE- an urban boundary layer meteorology project. Theor Appl Climatol 81:231–261

Roth M (2000) Review of atmospheric turbulence over cities. Q J Roy Meteorol Soc 126:941–990

Schatzmann M, Leitl B (2011) Issues with validation of urban flow and dispersion CFD models. J Wind Eng Ind Aerodyn 99:169–186

Schatzmann M, Bächlin W, Emeis S, Kühlwein J, Leitl B, Müller WJ, Schäfer K, Schlünzen H (2006) Development and validation of tools for the implementation of European Air Quality Policy in Germany (project VALIUM). Atmos Chem Phys 6:3077–3083

Schatzmann M, Olesen H, Franke J (eds) (2010) COST 732 model evaluation case studies: approaches and results. COST Office, Brussels, 121 pages. ISBN 3-00-018312-4

Tominaga Y, Mochida A, Yoshie R, Kataoka H, Nozu T, Yoshikawa M, Shirasawa T (2008) AIJ guidelines for practical applications of CFD to pedestrian wind environment around buildings. J Wind Eng Ind Aerodyn 96:1749–1761

Yoshie R, Mochida A, Tominaga Y, Kataoka H, Harimoto K, Nozu T, Shirasawa T (2007) Cooperative project for CFD prediction of pedestrian wind environment in the Architectural Institute of Japan. J Wind Eng Ind Aerodyn 95:1551–1578

Chapter 9
Wind Tunnel Experiment and Large Eddy Simulation of Pollutant/Thermal Dispersion in Non-isothermal Turbulent Boundary Layer

Ryuichiro Yoshie

Abstract This chapter firstly describes the necessity of validation study of CFD in relation to pollutant/thermal dispersion in urban areas by comparing CFD results with reliable wind tunnel experimental data. The second section explains a technique for simultaneously measuring fluctuating velocity, temperature, and concentration in non-isothermal turbulent layers. The third section introduces examples of pollutant/thermal dispersion experiments in non-isothermal turbulent boundary layers with different atmospheric stability conditions. This measurement technique was used for the wind tunnel experiments. The fourth section reviews various methods for generating inflow turbulence for large eddy simulation and shows some calculated results by large eddy simulation of pollutant/thermal dispersion that target the wind tunnel experiments mentioned above with the experimental data.

Keywords Non-isothermal flow • Atmospheric stability • Pollutant dispersion • Wind tunnel experiment • Large eddy simulation

9.1 Introduction

With recent progress in high-speed processing of personal computers and propagation of commercial CFD software, it is becoming possible to predict pedestrian wind environments around actual high-rise buildings by CFD. In order to assure the quality of the CFD simulations, an AIJ (Architectural Institute of Japan) CFD working group conducted many comparative and parametric studies on various building configurations (Yoshie et al. 2007a). Based on these validation studies, the AIJ published “Guidelines for practical applications of CFD to pedestrian wind environment around buildings” in 2007 and 2008 (AIJ 2007 (Japanese version),

R. Yoshie (✉)
Tokyo Polytechnic University, Kanagawa, Japan
e-mail: yoshie@arch.t-kougei.ac.jp

Y. Tamura, R. Yoshie (eds.), *Advanced Environmental Wind Engineering*,
DOI 10.1007/978-4-431-55912-2_9

Tominaga et al. 2008a (English version)). Since publication of the AIJ guidelines, application of CFD to environmental impact assessment of pedestrian wind for actual urban development is increasing in Japan.

The AIJ guidelines mainly targeted strong wind problems around high-rise buildings. Furthermore, rapid urbanization especially in Asian countries has been bringing about serious air pollution and urban heat island phenomena, which become more serious in weak wind conditions. The importance of urban ventilation in weak wind regions such as behind buildings and within street canyons is now broadly recognized as a countermeasure to the urban heat island phenomenon and air pollution problems. In such weak wind regions, buoyancy effect due to spatial temperature difference caused by solar and nocturnal radiations cannot be neglected. CFD is expected to become a useful tool to predict and assess these issues as well as strong wind problems. In order to apply CFD techniques to estimate ventilation and pollutant/thermal dispersion in urban areas, it is indispensable to assess calculation conditions and performance of turbulence models by comparing calculated results with experimental data. However, such validations of CFD for pollutant/thermal dispersion in non-isothermal turbulent boundary layers are quite rare in the wind engineering field. One reason is that few research institutes have wind tunnels that can control air temperature and surface temperature of ground (wind tunnel floor) and building models, and few reliable experimental data are available to validate CFD results. Wind velocity measurement by hot-wire anemometers in non-isothermal flows is difficult because output voltages from the hot wires are affected not only by wind velocity but also by temperature. Ohya measured wind velocity and temperature fluctuation simultaneously in a thermally stratified wind tunnel using an X-type hot wire and an I-type cold wire (Ohya 2001; Ohya and Uchida 2004). But it is difficult to apply their method to weak wind regions such as behind buildings and within street canyons where both positive and negative (reverse) flows exist. An X-type hot wire can measure wind velocity components only if no reverse flow occurs at any instantaneous moment. Furthermore, velocity, temperature, and concentration should be measured simultaneously to obtain turbulent heat fluxes and turbulent pollutant concentration fluxes, which are important turbulent statistics to validate CFD results. Laser Doppler velocimeter (LDV), which is not influenced by air temperature, may be used for simultaneous measurement. However, seeding particles for LDV measurement stuck to a cold wire or a thermocouple cause error in temperature measurement, and they can cause damage to instruments for pollutant concentration measurement. Thus, it is very difficult to measure wind velocity, temperature, and pollutant concentration simultaneously in non-isothermal turbulent boundary layers. This chapter describes a technique for simultaneously measuring fluctuating velocity, temperature, and concentration and introduces some results of validation studies on large eddy simulation of pollutant/thermal dispersion in non-isothermal boundary layers.

9.2 Technique for Simultaneously Measuring Fluctuating Velocity, Temperature, and Concentration

The authors (Yoshie et al. 2007b) have developed a system for simultaneously measuring fluctuating velocity, temperature, and concentration by refereeing and expanding Ohya's method (Ohya 2001; Ohya and Uchida 2004). This system is composed of a split film probe, a cold-wire thermometer, and a high-speed flame ionization detector and has the following characteristics:

1. Turbulent heat fluxes and turbulent concentration fluxes can be obtained from simultaneously measured instantaneous wind velocity, temperature, and concentration.
2. It is possible to distinguish between positive flow and negative (reverse) flow in weak wind regions by using a split film probe.
3. It achieves appropriate temperature compensation for the output voltage of the split film in a flow with a large temperature fluctuation.

9.2.1 *Calibrator of Split Film Probe and Cold Wire*

In order to precisely calibrate the split film probe and the cold wire under non-isothermal low wind speed condition, reference wind velocity and temperature and output voltages from the split film and the cold wire should be precisely measured under low turbulent conditions. Calibrators for the split film probe and the cold wire (Fig. 9.1) were developed for this purpose. A laminar flow meter precisely measures the flow volume rate inside the calibrator (diameter, D = 100 mm), and mean wind velocity in the calibration part, U_{mean}, is obtained by dividing the flow volume rate by the area of the cross section at the calibration part. Based on this U_{mean}, and by considering the wind velocity profile shown in Fig. 9.2 (discussed later), wind velocity around the center of the calibrator, U_c, is obtained. This U_c is the reference wind velocity used for calibration of the split film probe. A duct heater is used for temperature control, and the air temperature θ_a inside the calibrator is measured by a thermocouple with a wire diameter of 75 μm.

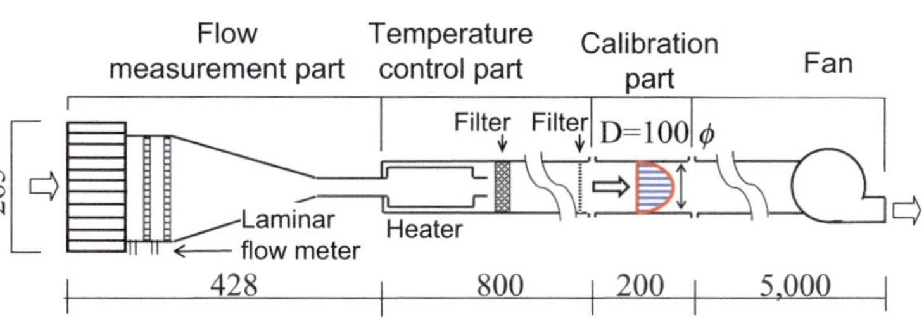

Fig. 9.1 Calibrator for split film probe and cold wire

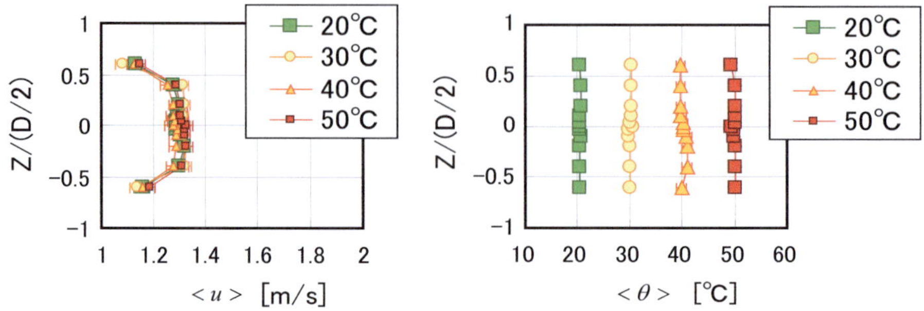

a) Distribution of wind velocity   b) Distribution of temeparure

Fig. 9.2 Distribution of wind velocity and temperature inside calibrator

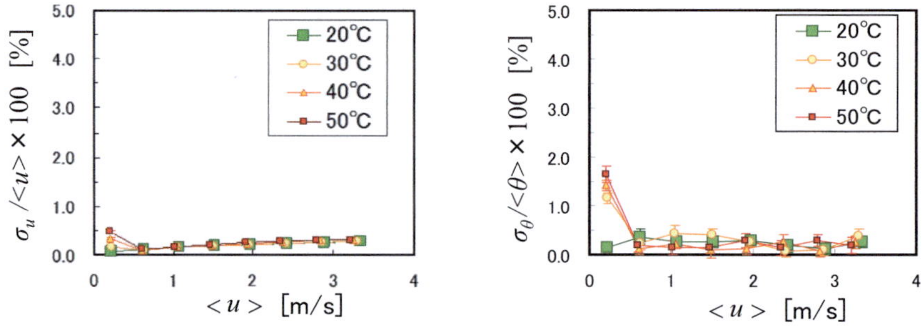

a) Turbulent intensity of wind velocity I_u b) Turbulent intensities of air temperature I_θ

Fig. 9.3 Turbulence intensities of wind velocity and temperature

This θ_a is the reference air temperature used for calibration of the cold wire and the split film probe. Insulating material is used for the outer package of the calibrator to reduce heat loss and control the temperature distribution inside the calibrator.

Figure 9.2 shows distributions of wind velocity and air temperature inside the calibrator. In the figures, $< >$ means time averaging, i.e., $<u>$ and $<\theta>$ are the time-averaged wind velocity and temperature, respectively. The distributions of wind velocity and air temperature are homogeneous around the center of the calibrator (Z/(D/2) = 0). Therefore, if the split film probe and the cold wire are placed around the center of the calibrator, error due to the sensor position is very small. Figure 9.3 shows the turbulence intensities of wind velocity, I_u, and turbulent intensities of air temperature, I_θ, at Z/(D/2) = 0. In the figure, σ_u and σ_θ mean standard deviations of wind velocity and temperature fluctuation, respectively. As shown, both turbulent intensities are very small: I_u<1 % and I_θ<2 %. Figure 9.4 shows the relative uncertainty (ISO 1993) of wind velocity using this calibrator, which is less than 1.5 %. However, when a wind tunnel is used instead of the calibrator, it is about 5.0 %. Thus, the calibrator greatly improves the reliability of the wind velocity calibration.

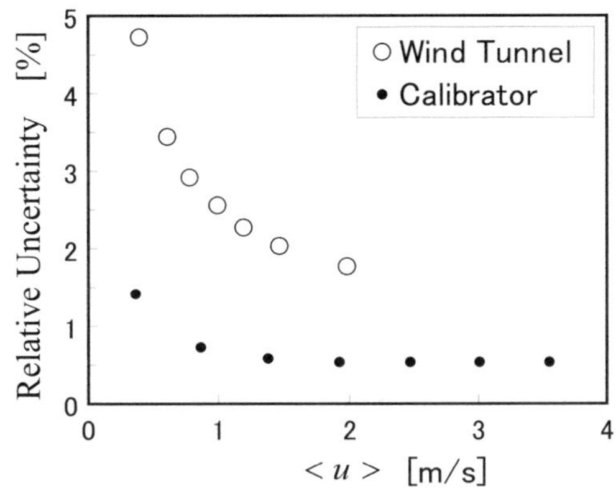

Fig. 9.4 Relative uncertainty of wind velocity measurement

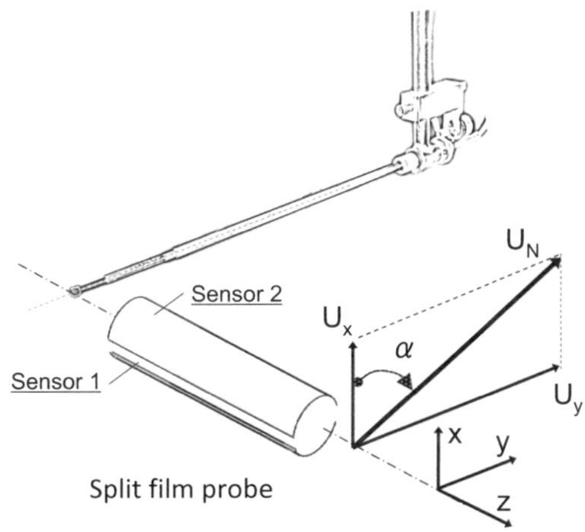

Fig. 9.5 Definition of coordinates of split film probe

9.2.2 Calibration Procedure for Cold Wire and Split Film Probe

In this measurement system, a split film probe of DANTEC (55R55 for u component, 55R57 for v component, 55R 56 for w component) and a CTA module (90C10) are adopted for wind velocity measurement; and a cold wire (55P31) and a temperature module (90C20) are used for temperature measurement. As shown in Fig. 9.5, the split film probe consists of two semicircular films which are split (insulated) from each other. The split film probe and the cold wire are placed at about 5 mm intervals in the calibrator so that the split film probe and the cold wire do not affect each other. The procedures for obtaining calibration data are as follows:

1. The angle of the wind approaching the split film probe is set as $\alpha = 0°$ (Fig. 9.5) under a constant air temperature, and 8 values of reference wind velocity U_c are

used. For each wind velocity, the output voltages E_1 and E_2 from sensor 1 and sensor 2 of the split film probe are measured, and reference air temperature θ_a by the thermocouple and output voltage E_{cw} from the cold wire are measured.
2. The angle of the wind approaching the split film probe is set as $\alpha = 180°$, and the above measurements are repeated.
3. The above procedures 1 and 2 are conducted for several reference air temperatures (e.g., 10–60 °C), to obtain E_1, E_2, and E_{cw} for each air temperature.

9.2.3 Calibration of Cold Wire

Figure 9.6 shows the relation between reference air temperature θ_a measured by the thermocouple and output voltage E_{cw} from the cold wire in the calibrator. As shown, E_{cw} changes linearly with θ_a and does not depend on wind velocity. Thus, the relation between E_{cw} and θ_a can be expressed as follows.

$$E_{CW} = A_C + B_C\theta_a \tag{9.1}$$

The calibration coefficients A_c and B_c (both constants) can be obtained by the least square method.

9.2.4 Measurement of Air Temperature in Wind Tunnel Experiment

In the wind tunnel experiments after the calibration procedure, instantaneous output voltage from the cold wire E_{cw} is measured and converted into instantaneous air temperature θ_a using the following equation, which is a deformation of Eq. (9.1).

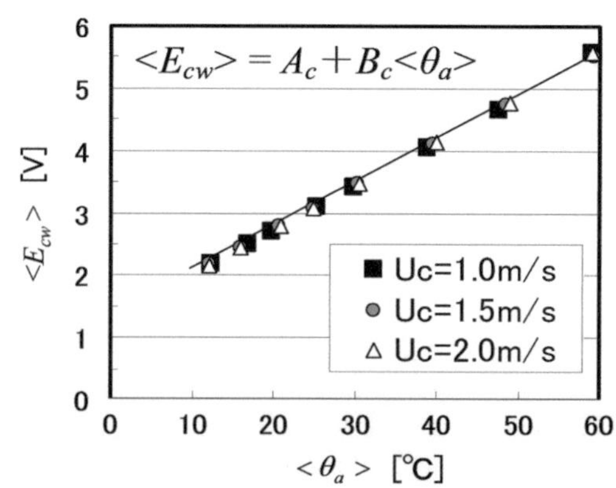

Fig. 9.6 Relation between reference air temperature θ_a and output voltage E_{cw} from cold wire

$$\theta_a = \frac{E_{cw} - A_c}{B_c} \tag{9.2}$$

9.2.5 Calibration of Split Film and Measurement of Wind Velocity Components in Wind Tunnel Experiment

Figure 9.7 shows the relation between reference air temperatures θ_a measured by a thermocouple and the sum of squares of output voltages $<E_1>^2 + <E_2>^2$ from the split film in the calibrator. The surface temperature of the split film is kept at a fixed high temperature θ_s. As the air temperature θ_a increases, the difference between θ_a and θ_s becomes smaller, and the convective heat transfer from the split film to air decreases, and so the output voltages E_1 and E_2 from the split film become lower. The relation in Fig. 9.7 can be expressed by the following equation.

$$< \theta_a > = \theta_s + C\left(< E_1 >^2 + < E_2 >^2\right) \tag{9.3}$$

Where C is a constant.

The intercept of this first-order approximation formula corresponds to the surface temperature of the split film θ_s, and it is obtained from the least square method.

The relation between output voltages E_1 and E_2 from the split film probe and the reference wind velocity for calibration U_c can be expressed by the following equations based on King's law.

$$\frac{< E_1 >^2 + < E_2 >^2}{\theta_s - < \theta_a >} = A_{(+)} + B_{(+)} < U_c >^m \tag{9.4}$$

$$\frac{< E_1 >^2 - < E_2 >^2}{\theta_s - < \theta_a >} = A_{(-)} + B_{(-)} < U_c >^m \tag{9.5}$$

In Figs. 9.8a and b, the vertical axes represent the left-hand sides of Eqs. 9.4 and 9.5, respectively, and the horizontal axes depict $<U_c>^m$. The value of m is specified

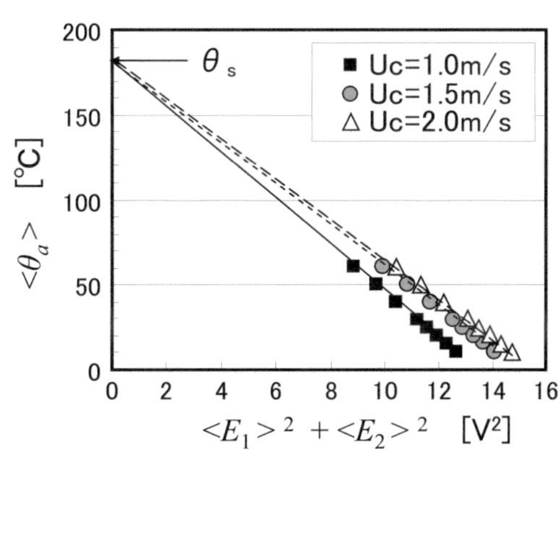

Fig. 9.7 Relation between reference air temperature and sum of squares of output voltages from split film probe

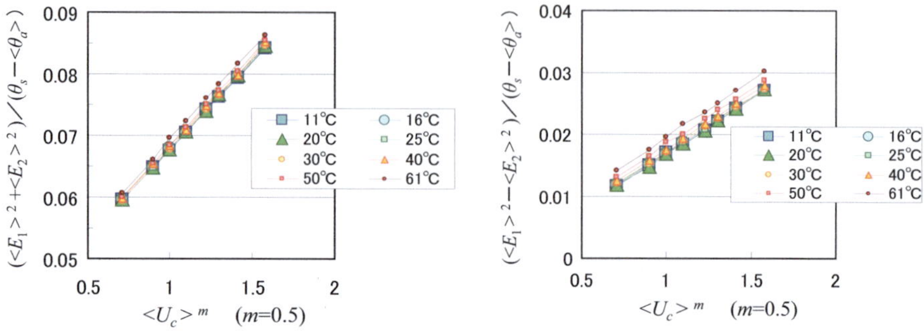

(a) Sum of squares of output voltages (b) Difference of squares of output voltages

Fig. 9.8 Relation between output voltages of split film and reference wind velocity

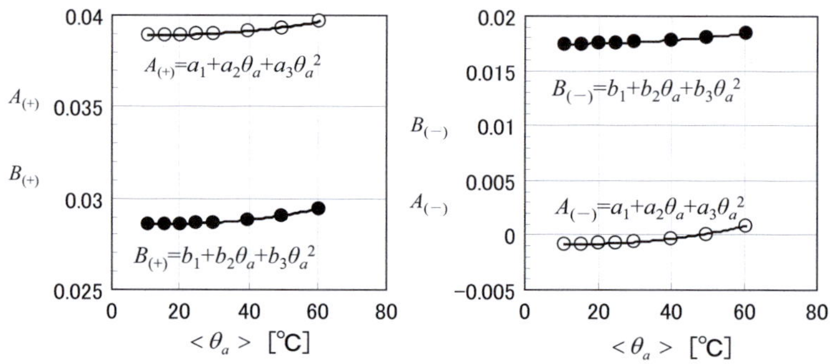

(a) $A_{(+)}$, $B_{(+)}$ (b) $A_{(-)}$, $B_{(-)}$

Fig. 9.9 Relation between calibration coefficients $A_{(+)}$, $B_{(+)}$, $A_{(-)}$, and $B_{(-)}$ and reference air temperature

so that the data can be best approximated by a linear function. In this calibration, $m = 0.5$. As shown in Fig. 9.8, since the calibration lines depend on air temperature θ_a, the calibration coefficients $A(_+)$, $B(_+)$, $A(_-)$, and $B(_-)$ become functions of θ_a, as shown in Fig. 9.9. Here, these coefficients are approximated by quadratic functions.

9.2.6 Measurement of Wind Velocity Components in Wind Tunnel Experiment

Figure 9.10 shows the relation between output voltages of the split film and the wind angle α to the split film (see Fig. 9.5).

In the wind tunnel experiments after the calibration procedure, calibration coefficients $A_{(+)}$, $B_{(+)}$, $A_{(-)}$, and $B_{(-)}$ for wind velocity are calculated from the instantaneous air temperature θ_a obtained from Eq. 9.2, and these calibration

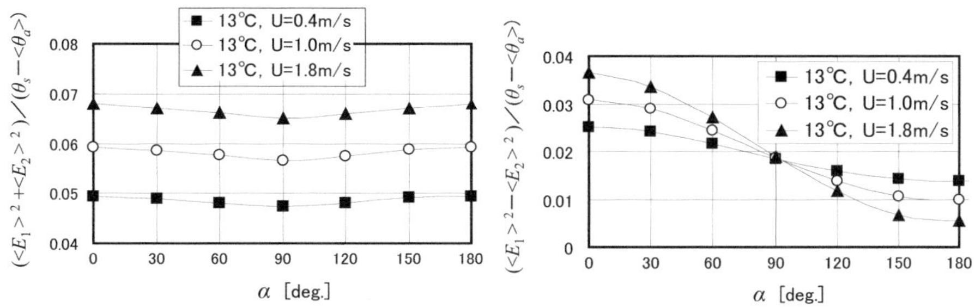

(a) Sum of squares of output voltages (b) Difference of squares of output voltages

Fig. 9.10 Relation between output voltages of split film and wind angle

coefficients are used in the following procedures. The scalar wind velocity u_N can be obtained from

$$\frac{{E_1}^2 + {E_2}^2}{\theta_s - \theta_a} = A_{(+)} + B_{(+)}{u_N}^m \tag{9.6}$$

As shown in Fig. 9.10b, the difference between the squared split film's output voltages C_α (Eq. (9.8)) can be approximated by a cosine curve, and so the instantaneous value of α is evaluated from Eq. 9.7:

$$\alpha = \begin{cases} \cos^{-1}\dfrac{C_\alpha - C_{90}}{|C_0 - C_{90}|} & (\text{for positive flow } (C_a - C_{90} \geqq 0)) \\ \cos^{-1}\dfrac{C_\alpha - C_{90}}{|C_{180} - C_{90}|} & (\text{for negative (reverse) flow } (C_a - C_{90} < 0)) \end{cases} . \tag{9.7}$$

where

$$C_\alpha = \frac{{E_1}^2 - {E_2}^2}{\theta_s - \theta_a} \quad \text{(vertical axes in Fig. 9.8b and Fig. 9.10b)} \tag{9.8}$$

$$C_0 = A_{(-),0} + B_{(-),0}{u_N}^m \tag{9.9}$$

where $A_{(-),0}$ and $B_{(-),0}$ are the calibration coefficients at $\alpha = 0°$.

$$C_{180} = A_{(-),180} + B_{(-),180}{u_N}^m \tag{9.10}$$

where $A_{(-),180}$ and $B_{(-),180}$ are the calibration coefficients at $\alpha = 180°$.

$$C_{90} = (C_0 + C_{180})/2 \tag{9.11}$$

The sum of squares of the split film's output voltages shown in Fig. 9.10a has variation about 5 % among the cases of α = 0°, 180°, and 90°, and so scalar wind velocity u_N obtained from Eq. 9.6 is somewhat affected by wind angle α. Therefore, U_N and α that satisfy both Eqs. 9.6 and 9.7 are calculated using an iteration method, and then U_N and α are determined.

Using the determined instantaneous values of U_N and α, the instantaneous wind velocity components u_x and u_y are obtained by the following equations.

$$u_x = u_N \cos\alpha \tag{9.12}$$

$$u_y = u_N \sin\alpha \tag{9.13}$$

Figure 9.11 compares the scalar wind velocity $<u_N>$ measured by this calibration method and that by an ultrasonic anemometer. In addition, Fig. 9.12 shows the measured value of $<C_\alpha - C_{90}>/<C_{0,180} - C_{90}>$ against the wind angle α, which is set by the angle controller. Both $<u_N>$ and α are measured precisely, and their relative uncertainties are below 5 %.

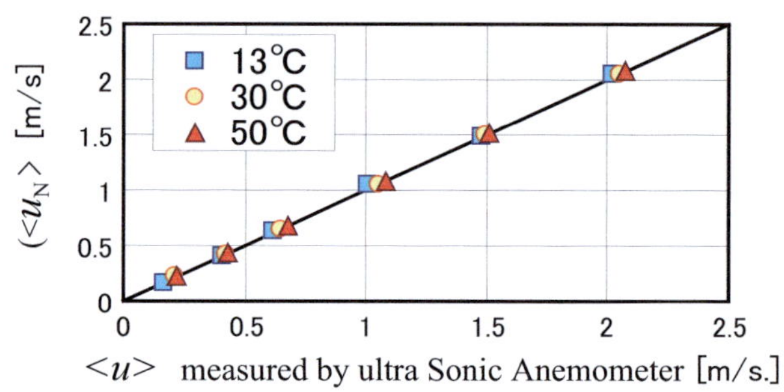

Fig. 9.11 Measurement accuracy of scalar wind velocity

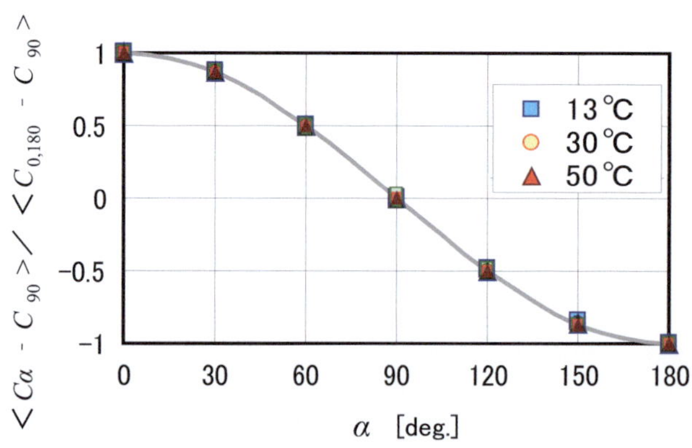

Fig. 9.12 Measurement accuracy of wind angle

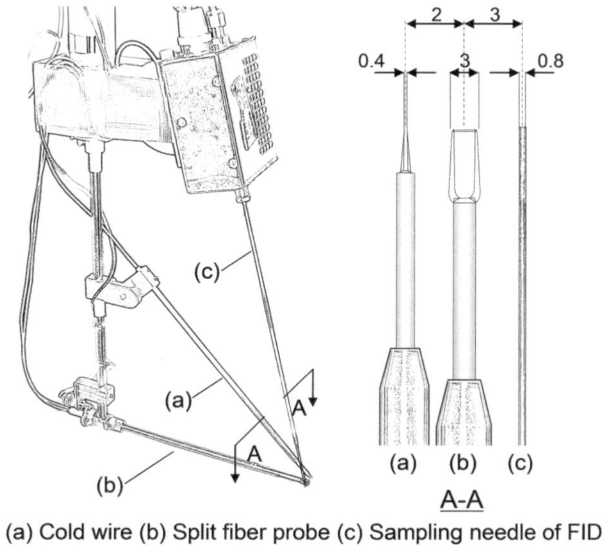

Fig. 9.13 Layout of sensors

9.2.7 *Layout of Sensors in Wind Tunnel Experiment*

In this system, it is necessary to place a split film, a cold wire, and a sampling tube of a high-speed flame ionization detector (FID) adjacent to one another. The influence of this layout on measured mean quantities and turbulent fluxes has been investigated (Yoshie et al. 2007b). Based on this investigation, the layout of sensors in the wind tunnel experiment was determined as shown in Fig. 9.13.

9.3 Examples of Wind Tunnel Experiments in Non-isothermal Turbulent Boundary Layer

9.3.1 *Thermally Stratified Wind Tunnel*

Figure 9.14 shows the thermally stratified wind tunnel in Tokyo Polytechnic University. This wind tunnel is a closed circuit type with a test section of 1.2 m wide, 1.0 m high, and 9.35 m long. The air is blown through a motor-driven fan of 5.5 kW DC. The maximum wind speed is 2.0 m/s at most because low wind speed is required to satisfy the similarity law of atmospheric stability condition.

The wind tunnel is equipped with a temperature profile cart (Fig. 9.15) and ambient air conditioners (Fig. 9.14). The air flow temperature can be controlled in the range of 10–60 °C by the temperature profile cart and ambient air conditioners. The temperature profile cart is installed between the contraction cone (contraction ratio = 0.25) and the test section. It has four aluminum screens (opening

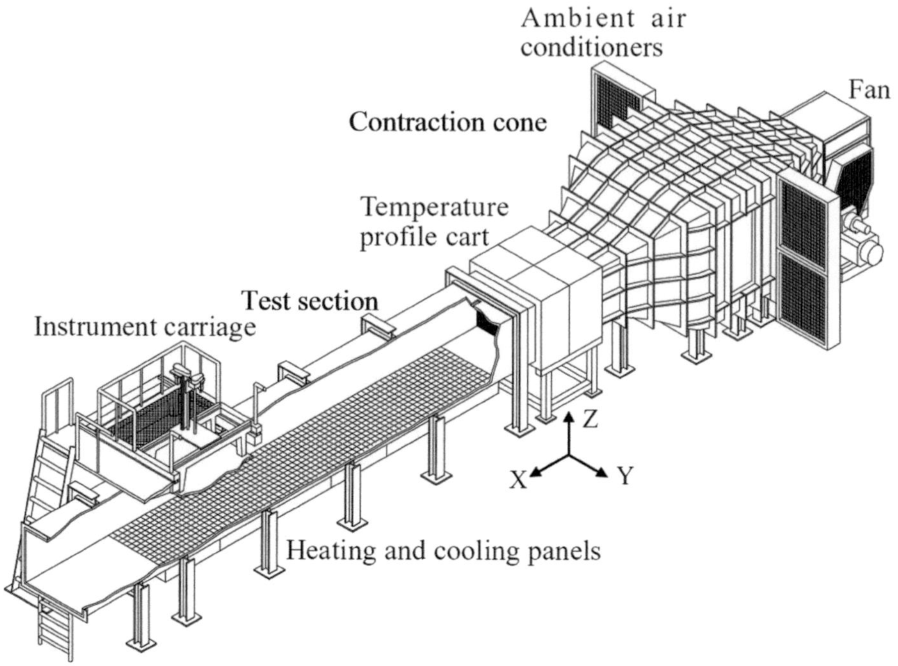

Fig. 9.14 Thermally stratified wind tunnel in Tokyo Polytechnic University

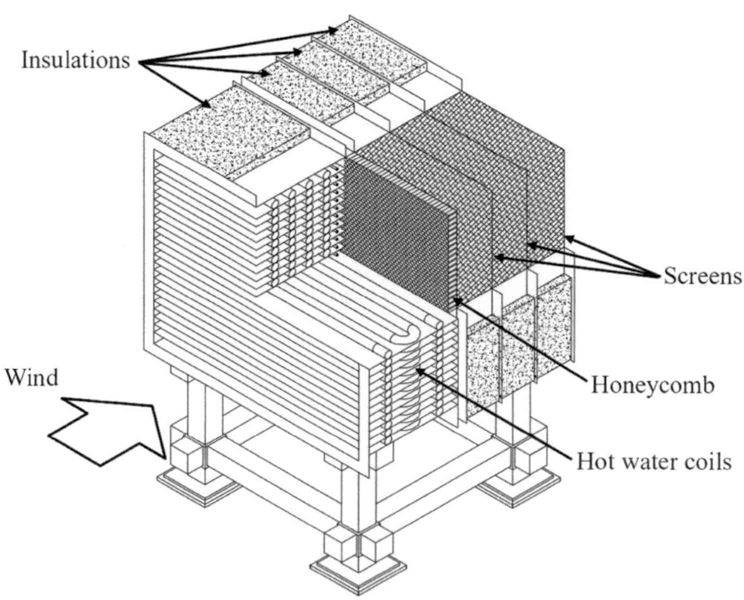

Fig. 9.15 Temperature profile cart

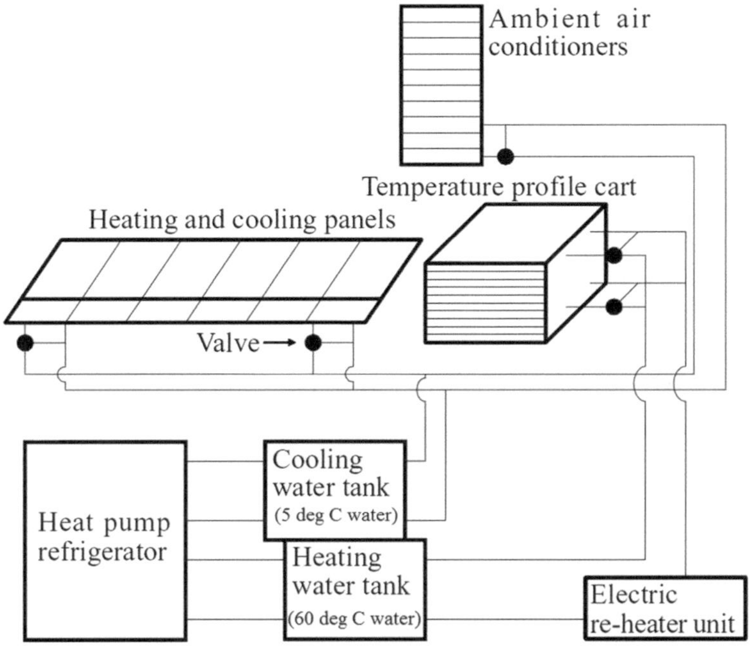

Fig. 9.16 Schematic diagram of temperature control systems

ratio = 61.5 %) at the diffuser section and a honeycomb after this section. The ambient air conditioners are responsible for maintaining the temperature outside the wind tunnel. This will maintain a uniform ambient temperature in the range of 8–20 °C. The floor of the test section consists of six heating and cooling panels with cooling coils and electric sheet heaters. The surface temperature of the floor can be controlled in the range between 10 °C and 80 °C.

Figure 9.16 is a schematic diagram of the temperature control systems of the wind tunnel. The temperature profile cart has hot water coils, and hot water is supplied from a heating water tank. This water is heated by an electric reheater unit. The ambient air conditioners also have water coils for which cold water is supplied from the cooling water tank. Both the hot water and the cold water are generated by heat pumps. The cold water in the tank is also used for the cooling panels of the wind tunnel floor.

9.3.2 *Wind Tunnel Experiment of Pollutant/Thermal Dispersion Behind a High-Rise Building*

As shown in Figs. 9.15, 9.16, and 9.17, wind velocity, temperature, and gas concentration around a high-rise building within turbulent boundary layers with

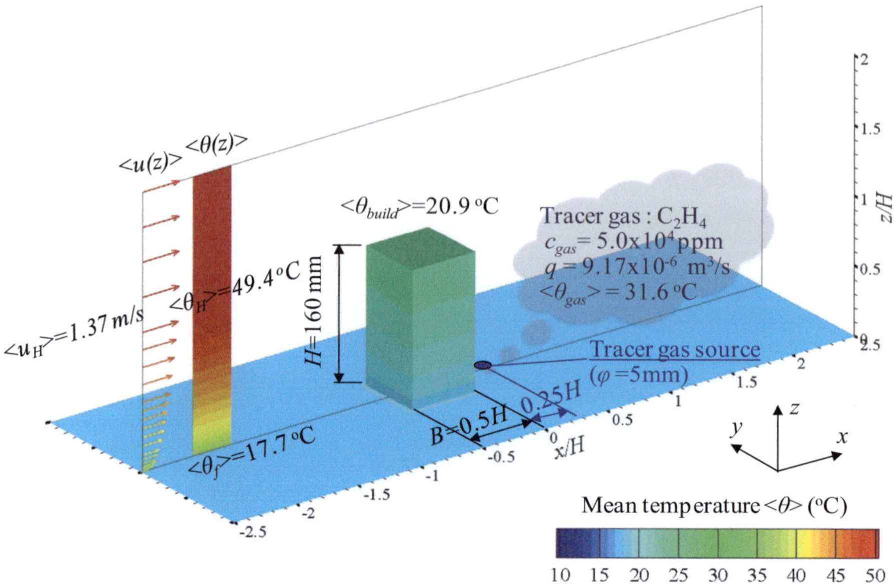

Fig. 9.17 Wind tunnel experiment on pollutant/thermal dispersion behind high-rise building within stable turbulent boundary layer

three different atmospheric stabilities (stable, neutral, and unstable) were measured using the technique described in Sect. 9.2. The purpose of this experiment was to produce data for CFD validation. The experiment was conducted in a thermally stratified wind tunnel in Tokyo Polytechnic University, shown in Fig. 9.14. The model building had a height (H) of 160 mm, a width (W) of 80 mm, and a depth (D) of 80 mm (H, W, $D = 2{:}1{:}1$) and was located in a turbulent boundary layer. The Reynolds number based on H (building height) and $<u_H>$ (approaching wind velocity at building height) was about 15,000. A point gas source was set on the floor 40 mm leeward of the model building. Tracer gas (C_2H_4, 5 % ethylene) was released from a hole (diameter, 5 mm) at a flow rate of $q = 9.17 \times 10^{-6}$ m^3/s. Table 9.1 summarizes the experimental conditions for the three atmospheric stability cases. The meanings of the symbols used in Figs. 9.17, 9.18, and 9.19 and Table 9.1 are described below.

$<\zeta>$	: time averaged value of ζ
ζ'	: fluctuation from time-averaged value of ζ, $\zeta' = \zeta - <\zeta>$
u, v, and w	: three components of wind velocity [m/s]
H	: building height (0.16 m)
u_H	: approaching wind velocity at building height [m/s]
θ	: air temperature [°C]
θ_f	: surface temperature of wind tunnel floor [°C]
θ_H	: air temperature of approaching wind at building height [°C]
$\Delta\theta$	: absolute value of temperature difference, $\Delta\theta = \mid \theta_H - \theta_f \mid$ [°C]

Table 9.1 Experimental Conditions

Case	(a) Stable condition	(b) Neutral condition	(c) Unstable condition
R_b	0.08	0.00	−0.10
H [m]	0.16	0.16	0.16
$<u_H>$ [m/s]	1.37	1.40	1.37
$<\theta_f>$ [°C]	17.7	21.2	45.3
$<\theta_H>$ [°C]	49.4	21.5	11.3
$<\Delta\theta>$ [°C]	31.6	0.4	33.9
$<\theta_0>$ [°C]	41.9	21.5	16.6
$<\theta_{build}>$ [°C]	20.9	21.1	41.7
$<\theta_{gas}>$ [°C]	31.6	21.2	30.4

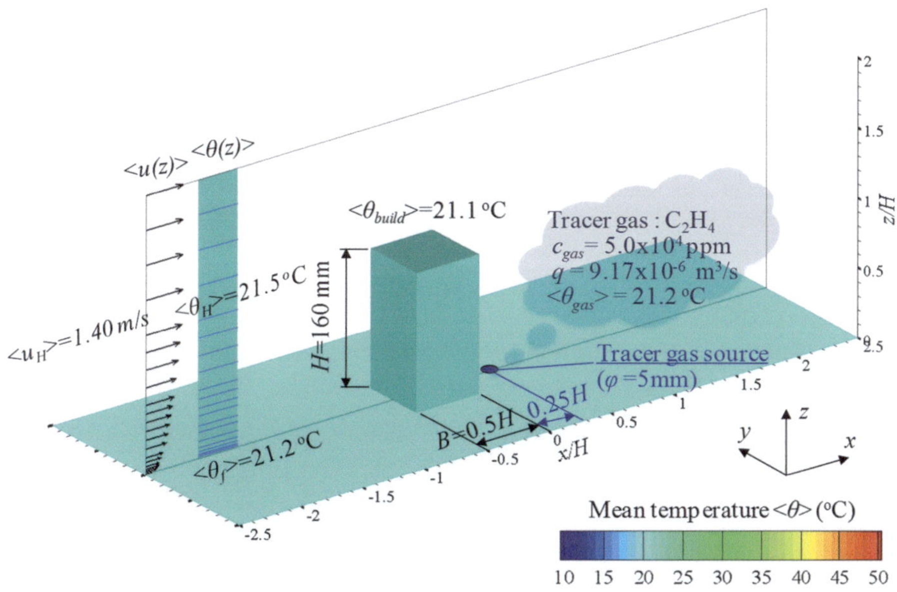

Fig. 9.18 Wind tunnel experiment on pollutant dispersion behind high-rise building within neutral turbulent boundary layer

θ_0	: space-averaged air temperature in boundary layer [°C]
θ_{gas}	: temperature of released tracer gas [°C]
θ_{build}	: surface temperature of building model [°C]
c	: gas concentration (ppm)
c_{gas}	: released tracer gas concentration (5.0×10^4 ppm)
q	: released gas emission rate (9.17×10^{-6} m^3/s)
c_0	: reference gas concentration, $c_0 = c_{gas}\, q / <u_H> H^2$ (ppm)
Ri_b	: Bulk Richardson number, $Ri_b = gH(<\theta_H> - <\theta_f>)/\{(<\theta_0> + 273) <u_H{}^2>\}$

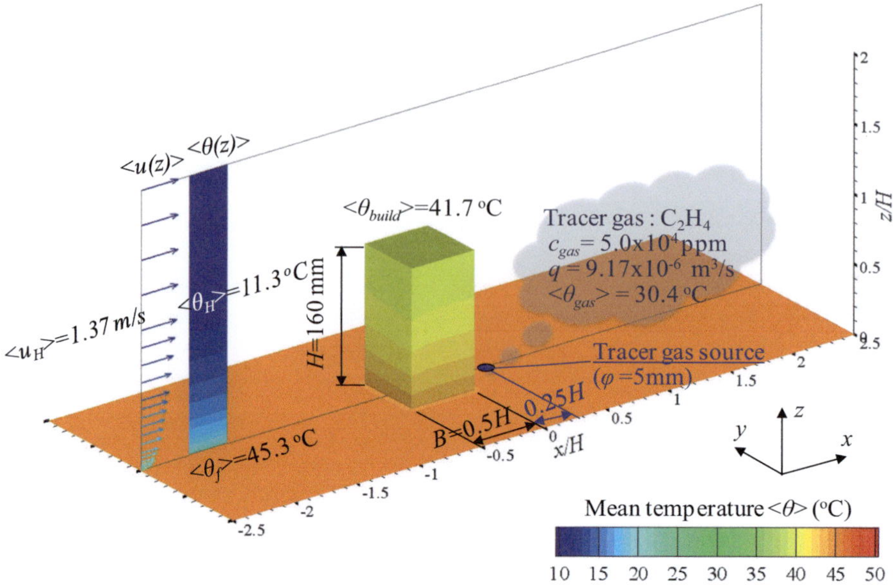

Fig. 9.19 Wind tunnel experiment on pollutant/thermal dispersion behind high-rise building within unstable turbulent boundary layer

The experimental database is open to the public on the website of Tokyo Polytechnic University (http://www.wind.arch.t-kougei.ac.jp/info_center/pollution/pollution.html).

The database includes the following quantities, which can be used for CFD validation.

- Mean wind velocity components: $<u>/<u_H>$, $<v>/<u_H>$, and $<w>/<u_H>$
- Mean scalar wind velocity: $<u_{sc}>/<u_H>$
- Normal stresses: $<u'^2>/<u_H{}^2>$, $<v'^2>/<u_H{}^2>$, and $<w'^2>/<u_H{}^2>$
- Shear stresses: $<u'v'>/<u_H{}^2>$ and $<u'w'>/<u_H{}^2>$
- Turbulent kinetic energy: $k/<u_H{}^2>$
- Mean temperature: $(<\theta> - <\theta_f>)/<\Delta\theta>$
- Standard deviation of temperature: $\sigma_\theta\ /<\Delta\theta>$
- Turbulent heat fluxes: $<u'\theta'>/(<u_H><\Delta\theta>)$, $<v'\theta'>/(<u_H>\ <\Delta\theta>)$, and $<w'\theta'>/(<u_H>\ <\Delta\theta>)$
- Mean concentration: $<c>/c_0$
- Standard deviation of concentration: σ_c/c_0
- Turbulent concentration fluxes: $<u'c'>/(<u_H>c_0)$, $<v'c'>/(<u_H>c_0)$, and $<w'c'>/(<u_H>c_0)$

9.3.3 Wind Tunnel Experiment on Pollutant/Thermal Dispersion Within Building Block

The second example is a wind tunnel experiment on pollutant/thermal dispersion within a building block (Yoshie and Hu 2013). Figure 9.20 shows the experimental setup. This experiment was also conducted in a thermally stratified wind tunnel in Tokyo Polytechnic University shown in Fig. 9.14. The turbulent boundary layers were generated by 26 very thin roughness elements made of thin aluminum plates located upstream. They created a long, rough upwind fetch to generate a turbulent boundary layer.

A total of $9 \times 14 = 126$ cubic blocks was put in the turbulent boundary layer located downstream (Figs. 9.20 and 9.21) to represent building blocks. Each building block had the same configuration: L (60 mm) × W (60 mm) × D (60 mm). The city blocks were spaced 60 mm apart in both the x and y directions. Figure 9.21 shows an overview of the building block arrangement. Tracer gas ethylene (C_2H_4) was released from a line of the floor. Figure 9.22 shows measuring points. The locations of the measuring points were selected so that various flow patterns (reverse flow, upward flow, and downward flow in the street canyons and flow on the roads) were included.

The purpose of this wind tunnel experiment was to produce data for CFD validation and also to investigate the effect of atmospheric stability on pollutant concentration in a city. Thus, atmospheric stability was changed in five cases as shown in Table 9.2. The reference height H_R was 0.32 m, and the velocity at this height of inflow boundary was set as reference velocity (U_R). The atmospheric stability was characterized by Bulk Richardson number (Ri_b). Bulk Richardson number can be expressed as follows:

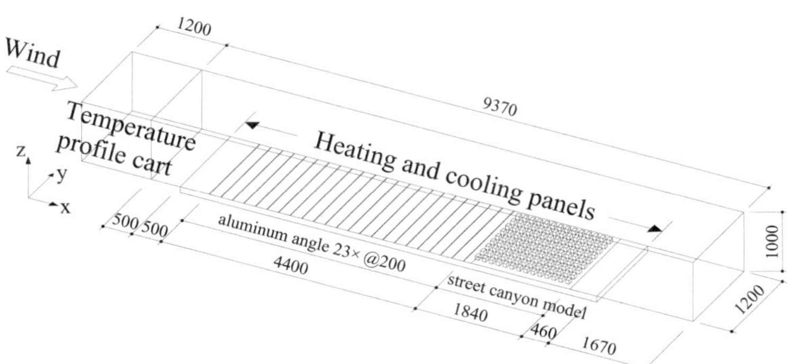

Fig. 9.20 Experimental setup (units, mm)

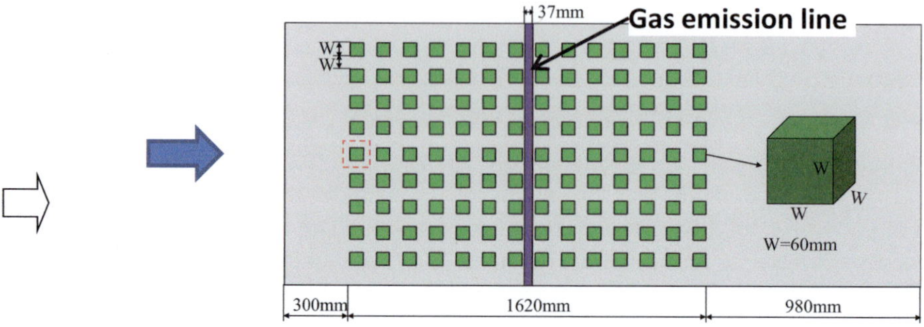

Fig. 9.21 Arrangement of building blocks

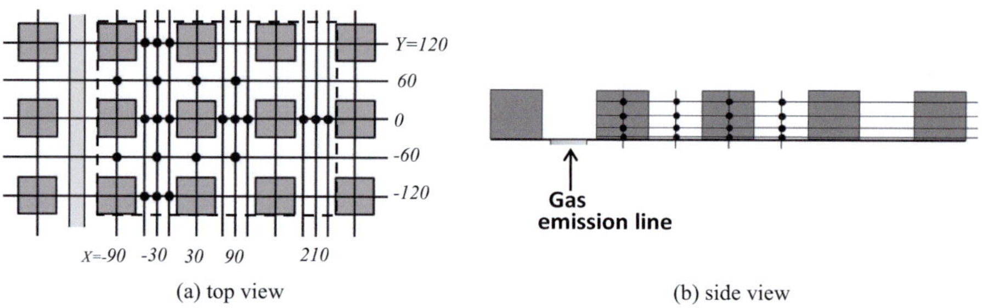

Fig. 9.22 Measuring points

Table 9.2 Atmospheric stability conditions

	Unstable	Weakly unstable	Neutral	Weakly stable	Stable
U_R [m/s]	1.4	1.8	1.8	1.5	1.2
θ_R [°C]	9	10	11	48	55
θ_f [°C]	49	41	11	16	14
$\Delta\theta$ [°C]	40	31	0	32	41
θ_0 [°C]	13	14	11	48	55
Ri_b	−0.23	−0.1	0	0.13	0.29

$$Ri_b = \frac{gH_R \times (\theta_R - \theta_f)}{(\theta_0 + 273) \times U_R^2}$$

where g is the acceleration due to gravity [m/s^2], H_R is the reference height [m], θ_R is the temperature at reference height [°C], θ_f is the surface temperature of the wind tunnel floor [°C], T_0 is the average inflow temperature [°C], and U_R is the velocity at reference height [m/s].

Ri_b for 5 inflow profiles are summarized in the last row of Table 9.2. Values of Ri_b ranged from −0.23 (unstable) to 0.29 (stable).

Figure 9.21 shows correlations for normalized nondimensional concentration C^* between neutral and non-neutral conditions. The normalized nondimensional concentration C^* is defined as follows:

$$C^* = \frac{CU_R H_R^2}{q}$$

where C is concentration [−] and q is emission rate of tracer gas [m^3/s].

As shown in Fig. 9.23, data are plotted almost on a single straight line. Thus, the ratio between C^* under non-neutral and neutral conditions were almost independent of the measurement locations. The slope of the line becomes larger with increase in Bulk Richardson's number Ri_b. C^* under unstable conditions were smaller than C^* under neutral condition, and C^* under stable conditions were larger than C^* under neutral condition.

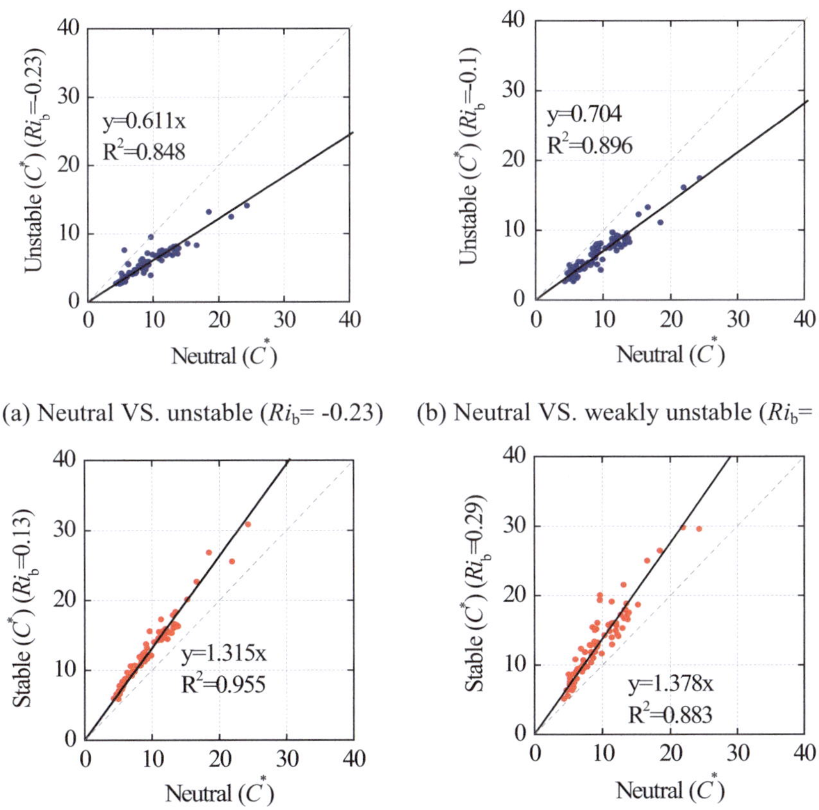

(a) Neutral VS. unstable (Ri_b= -0.23) (b) Neutral VS. weakly unstable (Ri_b= -0.1)

(c) Neutral VS. weakly stable (Ri_b= 0.13) (d) Neutral VS. stable (Ri_b= 0.29)

Fig. 9.23 Correlations of C^* between neutral condition and non-neutral conditions

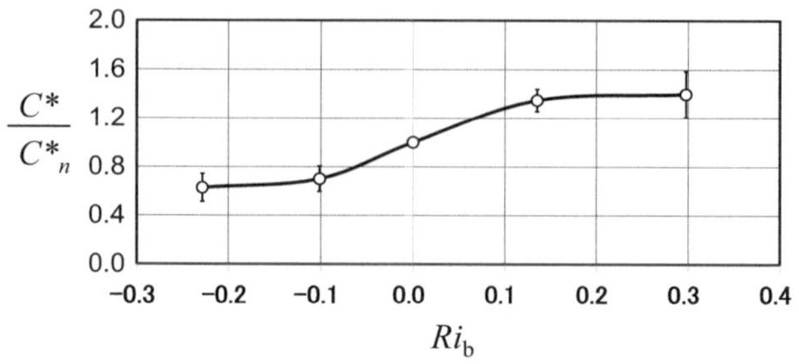

Fig. 9.24 Stability effect ratio (SER) = ratio between C* under non-neutral condition and C^*_n under neutral condition

Figure 9.24 shows the ratio between C^* under non-neutral conditions and C^*_n under neutral condition obtained from the experiment. We call this stability effect ratio (SER) on pollutant concentration (SER = C^*/C^*_n). Averaged SER ± σ (standard deviation) of all the measuring points are plotted in the figure. As shown, with increase in Ri_b, the SER increases. Since the standard deviation is relatively small, the SER is almost independent of location. But further case studies are necessary with different configurations of the city model to confirm the generality of the SER. If the function of the SER is universal, we can predict pollutant concentration under non-neutral conditions from experimental results under neutral conditions by using the function.

9.4 Large Eddy Simulation of Pollutant/Thermal Dispersion in Non-isothermal Turbulent Boundary Layer

9.4.1 Generation of Inflow Turbulence for Large Eddy Simulation

Turbulent flows are complex states of fluid motion and are characterized by eddies with a wide range of length and time scales. When simulating turbulent atmospheric boundary layers using large eddy simulation (LES), a crucial issue is how to impose physically correct turbulence at the inflow boundary of LES. The incoming flow should have these spatial and temporal characteristics. The influence of inflow turbulence on large eddy simulation has been presented by Tominaga et al. (2008b). It has been confirmed that the inflow turbulence for LES is extremely important. Several techniques have been proposed for generating inflow turbulence for LES in a neutral boundary layer. Most of them can be classified into four basic categories: random noise (white noise), synthetic method, precursor simulation, and recycling method.

The simplest way is to superimpose random fluctuations on the mean velocity profile with amplitude determined by the turbulence intensity level. But due to the lack of the required characteristics of turbulent flow, random noise is not an appropriate inlet condition.

One commonly used way is the artificial generation method (synthetic method) in which velocity fluctuations are obtained from an inverse Fourier transform of a prescribed spectrum that satisfies the characteristics of power spectral density and spatial correlations of the turbulent boundary layer. A method of generating the inflow turbulence for computational wind engineering applications based on this approach was developed by Kondo et al. (1997), Maruyama et al. (1999), and Iizuka et al. (1999).

A digital-filter-based generation of turbulent inflow conditions was presented by Xie and Castro (2008), and the method was validated by simulating plane channel flows with smooth walls and flows over arrays of staggered cubes (a generic urban-type flow). Another synthetic method for generating realistic inflow conditions was presented by Druault et al. (2004) based on proper orthogonal decomposition (POD) and linear stochastic estimation (LSE). But its spatial resolution is not sufficient due to the limited number of hot-wire probes that simultaneously measure velocity. Perret et al. (2006) took the same approach based on the use of POD, but they coupled it with a database obtained by stereoscopic particle image velocimetry (SPIV).

Another method that has been used to generate inflow conditions involves running a separate precursor calculation of an equilibrium flow to generate a library of turbulent data that can be introduced into the main computation at the inlet (Tabor and Baha-Ahmadi 2010). This has the advantage that the inflow conditions for the main computation are taken from a genuine simulation of turbulence and thus should possess many of the required characteristics, including temporal and spatial fluctuation with correlation and a correct energy spectrum.

Instead of simulating the entire upstream region using precursor simulation, the most commonly used way is to employ a recycling method in which the velocity in a downstream plane is used (recycled) for the inflow boundary. In this method, the region used to generate turbulence is usually referred to as "driver region," and the second region that we are interested in is usually referred to as "main region." Lund et al. (1998) first proposed this rescaling recycling method to generate developing turbulent inflow data for LES. Kataoka and Mizuno (2002) simplified Lund's method by assuming that the boundary layer thickness is constant within the driver section. Nozawa and Tamura (2002, 2005) extended Lund's method to a rough-wall boundary layer flow using a roughness block arrangement. Now these methods are widely used to generate inflow turbulence for LES in neutral boundary layers.

The non-isothermal boundary layer (unstable or stable) is a very common atmospheric phenomenon, but there have been few LES studies on stratified atmospheric flows. When LES is applied to a non-isothermal field, not only inflow velocity fluctuation but also temperature fluctuation is necessary. Kong et al. (2000) proposed a method for generating inflow temperature fluctuation with reference to Lund's method according to the similarity between temperature and stream-wise velocity. To the author's knowledge, this is the first time inflow turbulence in the non-isothermal condition has been dealt with. In this method, the velocity fluctuation was generated using Lund's method, and the inflow temperature fluctuation was generated using the same rescaling and recycling law as that used for streamwise velocity. Hattori et al. (2007) conducted an unstable and stable boundary layer simulation, and the turbulent inflow data were generated by this method (Lund et al. 1998; Kong et al. 2000). In their calculation, the bulk Richardson number was −0.01 for the unstable condition and 0.01 for the stable condition, which corresponds to very weak thermal stratification. In the driver region, the neutral boundary layer was simulated, and the temperature was treated as a passive scalar.

Tamura et al. (2003) proposed a method for dealing with the thermally stratified effect. In the driver region, velocity fluctuation was generated using the quasi-periodic boundary condition for a rough wall, while temperature was treated as a passive scalar, and a mean temperature profile was given to the inflow condition of the driver region. Generated inflow data for temperature as well as velocity were introduced into the main computation domain, where the solution of physical quantities took into account buoyancy effects. But they also pointed out that fluctuation of the passive scalar was too large and not appropriate for the inflow condition of a stable turbulent boundary layer. Therefore, for stable conditions, the generated inflow velocity data at the recycling station were introduced into the main computational region, but a mean temperature profile without fluctuation was imposed at the inflow boundary of the main region (Tamura 2008). In the simulation of the urban heat island phenomena of the downtown region of Tokyo, as the flow coming from the sea was cold air, they used two driver regions to generate inflow turbulence for LES (Tamura et al. 2006). The driver region in domain 1 generated a neutral turbulent boundary layer by using the rescaling technique, and domain 2 thermally stabilized the turbulent boundary layer based on the sea breeze characteristics. Then the generated data were used for domain 3 to simulate the urban heat island phenomena of Tokyo. Abe et al. (2008) conducted LES calculation of gas dispersion in a convective boundary layer (CBL) and investigated the characteristics of turbulence structures and gas dispersion behavior. The inflow velocity and temperature fluctuations were generated using the method described by Tamura et al. (2003) for unstable boundary layers.

Brillant et al. (2008) also developed a thermal turbulent inflow condition based on parallel flows in order to simulate a turbulent thermal boundary layer. Then they tested this thermal turbulent inflow condition through a turbulent plane channel simulation. In this simulation, when velocity and temperature inflow fluctuations were given simultaneously, the fluctuating temperature profiles were well maintained as the flow proceeded downstream. If velocity inflow fluctuation and only a mean temperature profile (no fluctuation) were given as inflow condition, the temperature fluctuation gradually developed as the flow proceeded downstream. But the turbulent velocity field did not quickly generate thermal fluctuations.

Yoshie et al. (2011) used a precursor method to generate velocity and temperature fluctuations simultaneously in a non-isothermal boundary layer. In this method, the total simulations were composed of two domains: a pre-simulation domain to generate inflow velocity and temperature fluctuations for LES and a main domain to simulate gas dispersion around a high-rise building in a non-isothermal boundary layer. In the pre-simulation, the whole wind tunnel and all the aluminum plates (roughness elements) were reproduced by large eddy simulation using a buoyant solver. As a continuation of the study, which only focused on an unstable case using a precursor method, Yoshie et al. (2011) and Jiang et al. (2012) investigated two inflow turbulence generation methods (precursor method and recycling method) in both unstable and stable boundary layers.

Okaze and Mochida (2014) proposed a new method of generating turbulent fluctuations of wind velocity and scalar parameters such as temperature and pollutant concentration based on the Cholesky decomposition of Reynolds stresses and turbulent scalar fluxes by expanding the methods of Xie and Castro (2008).

However, these kinds of researches on large eddy simulation of non-isothermal turbulent boundary layers are still quite rare, and methods of generating temperature fluctuation have not been sufficiently examined yet. Further investigations are anticipated.

9.4.2 Validation of Large Eddy Simulation of Pollutant/Thermal Dispersion in Non-isothermal Boundary Layer

This section introduces some of the validation studies on large eddy simulation of pollutant/thermal dispersion in an unstable boundary layer conducted by the authors.

9.4.2.1 Generation of Inflow Turbulence of Wind Velocity and Temperature for Large Eddy Simulation

Wind tunnel experiments described in Sects. 9.3.2 and 9.3.3 used the same experimental setup for producing a turbulent boundary layer, i.e., 26 very thin aluminum plates 9 mm high placed at L = 200 mm pitch on the wind tunnel floor as shown in Fig. 9.25.

Two turbulence generation techniques (precursor method and recycling method) were examined to reproduce the turbulent boundary layer shown in Fig. 9.25 (Jiang et al. 2012). Then the generated turbulence data were used for the inflow boundary condition for LES of gas/thermal dispersion behind a high-rise building (see Sect. 9.3.2) and within a street canyon (see Sect. 9.3.3) in an unstable turbulent boundary layer.

Firstly, a method for generating both velocity and temperature fluctuations simultaneously in non-isothermal boundary layers by precursor simulation is discussed here. In a previous study by Ohya and Uchida (2008), they simulated a

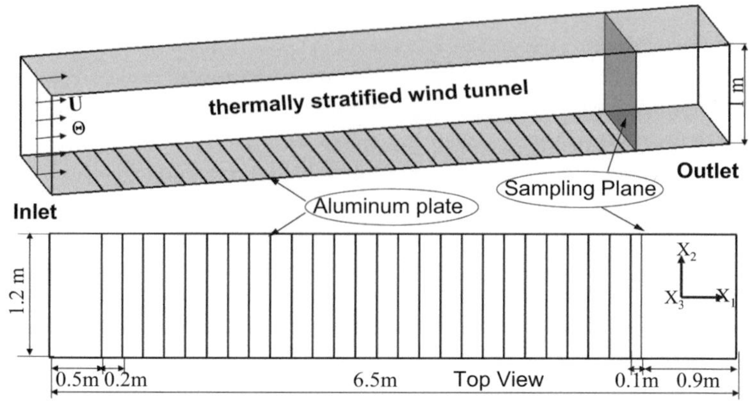

Fig. 9.25 Experimental setup for producing turbulent boundary layer (Jiang et al. 2012)

very long atmospheric boundary layer and investigated the influence of thermal stability on turbulence. We also simulated a long spatially developing turbulent boundary layer (including the transition from laminar flow to turbulent flow) but focused on the purpose of generating inflow turbulence for LES. In the precursor simulation, the whole wind tunnel and all the aluminum plates (Fig. 9.25) were reproduced by LES using a buoyant solver. The plates were treated as having zero thickness in the simulation. The wind velocity and temperature distribution at the inlet of the wind tunnel were spatially uniform, and turbulent intensity was very small (less than 1 %), so a uniform velocity $U = 1.43$ m/s and a uniform temperature $\Theta = 9.4$ °C without turbulence were given to the inflow boundary of the pre-simulation. A zero gradient condition was used for the outlet boundary condition. A no-slip boundary condition was applied to the wall shear stress on the floor. The nondimensional distances from the surfaces to the first fluid cells were below 1.0 for most regions. As thermal boundary conditions, the surface temperature was 45.3 °C, and a heat conduction boundary condition (Fourier's law) was applied for the heat flux on the floor surface. The sampling plane to obtain fluctuating velocity and temperature data was set at 0.1 m (11 times the aluminum plate height) downstream of the last aluminum plate.

Another method for generating inflow turbulence in a non-isothermal boundary layer using a recycling procedure was also investigated here (Fig. 9.26). Tamura et al. (2003) proposed a method for dealing with the thermally stratified effect. In the driver region, velocity fluctuation was generated using Lund's method (Lund et al. 1998) for a rough wall, while temperature was treated as a passive scalar, and a mean temperature profile was given to the inflow condition of the driver region. The same concept was adopted here, but the velocity fluctuation was generated using Kataoka's method (Kataoka and Mizuno 2002) with the roughness ground arrangement described by Nozawa and Tamura (2002). The roughness elements were exactly the same as those used in the wind tunnel experiment, but a short domain was adopted here. A mean velocity profile that came from the experimental

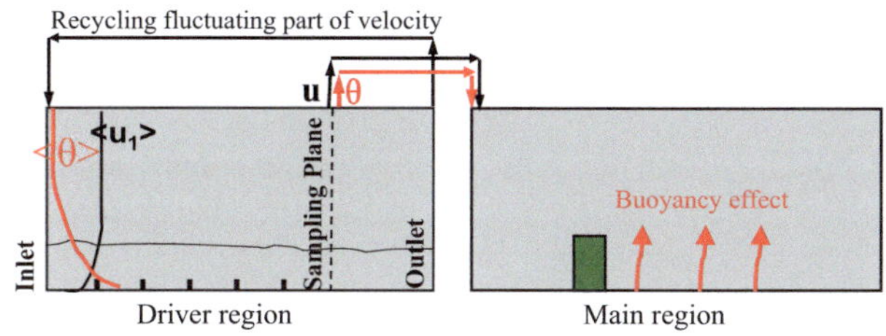

Fig. 9.26 Inflow turbulence generation by recycling method

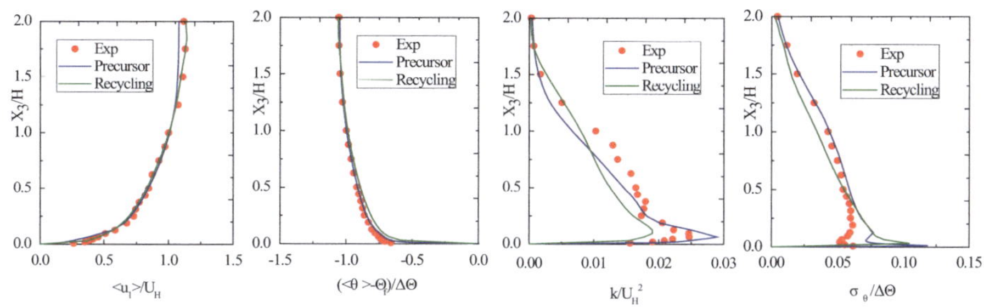

Fig. 9.27 Properties of generated turbulent flow by precursor and recycling methods

measurement was prescribed for the inflow condition, and only the fluctuating part was recycled between outlet station and inlet station. The following damping function (Kataoka 2008) was used to restrain development of the velocity fluctuation:

$$\varphi(\eta) = \frac{1}{2}\left\{1 - \tanh\left[\frac{8.0(\eta - 1.0)}{-0.4\eta + 0.82}\right]/\tanh(8.0)\right\} \tag{9.14}$$

where $\eta = z/\delta$. δ is the boundary layer thickness (0.25 m). Only a neutral boundary layer (NBL) was simulated in the driver region, the temperature was treated as a passive scalar, a mean temperature profile of the experiment was prescribed at the inflow boundary of the driver section, and we tried to use the fluctuating velocity field to generate a fluctuating temperature field.

Figure 9.27 shows the properties of generated flow by both the precursor method and the recycling method in the sampling position. The mean wind velocity, mean temperature, turbulent kinetic energy, and the r.m.s. value of temperature fluctuation agreed well with those of the experiment. Both methods can be used to generate turbulent inflow data for LES in an unstable boundary layer.

9.4.2.2 Examples of Large Eddy Simulation of Gas/Thermal Dispersion in Unstable Turbulent Boundary Layer

The generated inflow turbulence data were saved in a hard disk and used for an inflow boundary condition of LES for gas/thermal dispersion behind a single building (see Sect. 9.3.2) and within street canyons (see Sect. 9.3.3).

Firstly, LES results and experimental data of gas/thermal dispersion behind a single building within unstable turbulent boundary layer are compared (Yoshie et al. 2011).

Mean stream-wise velocities $<u_1>$ of the experiment and the calculations are shown in Fig. 9.28. The results from the RANS model (two-equation heat-transfer model (Nagano and Kim 1988)) show overestimation of the recirculation size behind the building, and the calculated downward flow in the region around $X_1/H = 0.7–1.5$ is weaker than the experimental one. The calculated reverse flow near the ground and the rising flow along the rear surface of the building are stronger than those of the experiment. On the other hand, the recirculation size by LES is narrower and the reverse flow near the ground and the rising flow along the rear surface of the building are weaker than those of the RANS model, which is closer to the experimental results.

Figures 9.29 and 9.30 show the distributions of mean gas concentration. In the RANS calculation result, the high concentration area near the ground does not spread downwind of the gas emission point (marked as a "black triangle"). Calculated gas concentration along the rear surface of the building was higher due to the rising flow from the ground. The calculation does not reproduce the periodic

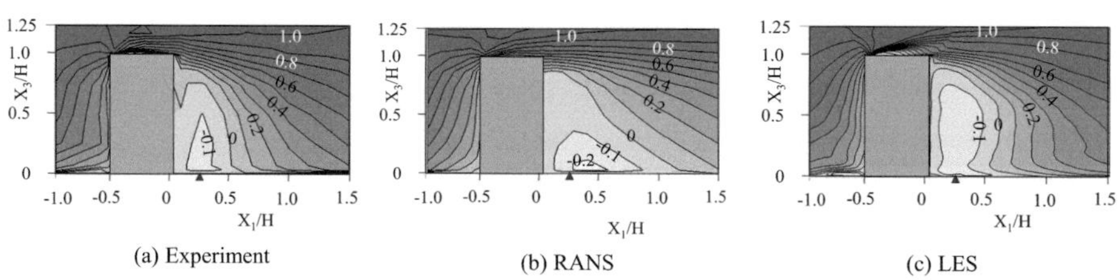

Fig. 9.28 Vertical distribution of mean velocity $<u_1>/U_H$ (Yoshie et al. 2011)

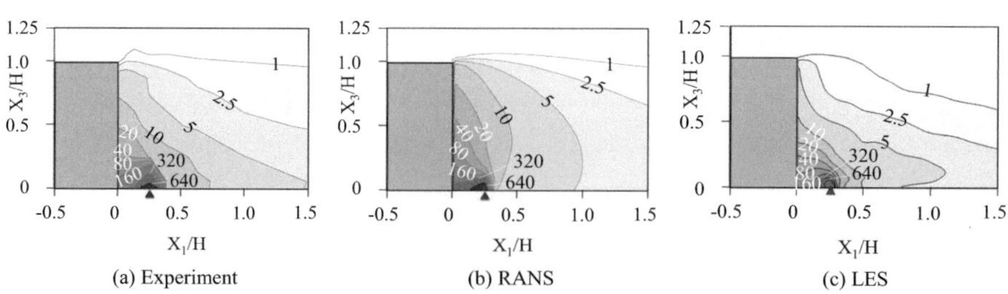

Fig. 9.29 Vertical distribution of mean concentration $<c>/C_0$ (Yoshie et al. 2011)

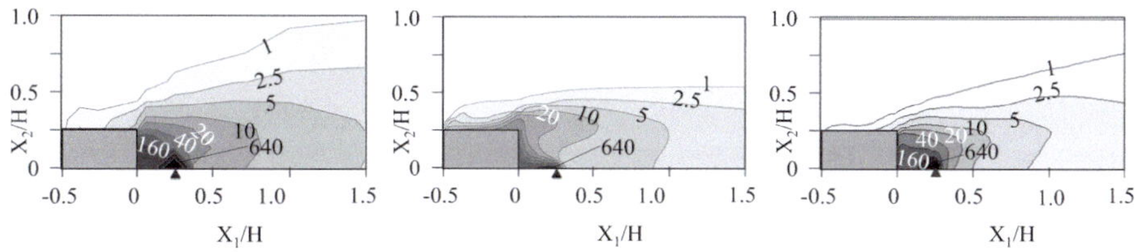

(a) Experiment (b) RANS (c) LES

Fig. 9.30 Horizontal distribution of mean concentration $<c>/C_0$ (Yoshie et al. 2011)

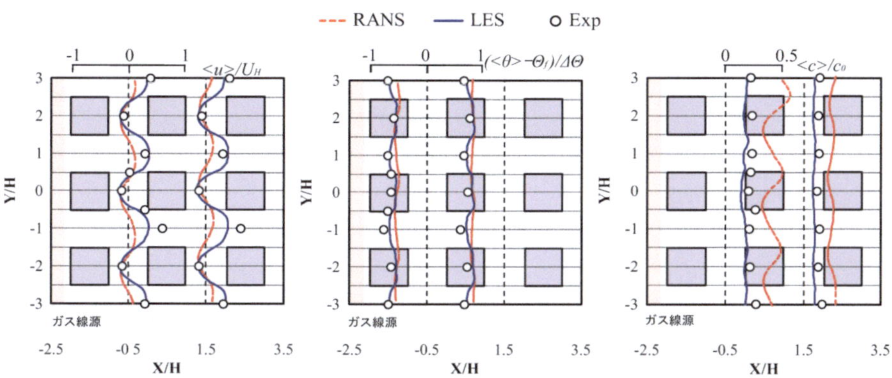

(a) Mean streamwise velocity (b) Mean temperature (c) Mean concentration

Fig. 9.31 Comparison between experimental result and CFD results ($z = H/6$)

fluctuations due to vortex shedding, and as a result, dispersion in the X_2 direction (lateral direction) is inhibited (Fig. 9.30). The distribution pattern of the LES is much closer to that of the experiment than of the RANS model, especially regarding the lateral width of gas dispersion.

Secondary, large eddy simulation of pollutant/thermal dispersion within a building block in an unstable turbulent boundary layer which targets the wind tunnel experiment described in Sect. 9.3.3. "Unstable" case in Table 9.2 was calculated by LES, and the results were compared with the experimental data.

Figure 9.31 shows the distributions of mean stream-wise velocity, mean temperature, and mean concentration near the floor (z = H/6 and H is building height). The LES results agreed well with the experimental data especially for the mean concentration. The RANS model (with standard k-ε model) overestimated the concentration by about 200 %. This is because the intermittent air exchange between street canyon and upper atmosphere cannot be reproduced in RANS, while it is well captured by LES. Figure 9.32 shows the correlations of mean concentration between experiment and LES results. The LES results are very close to the experimental data.

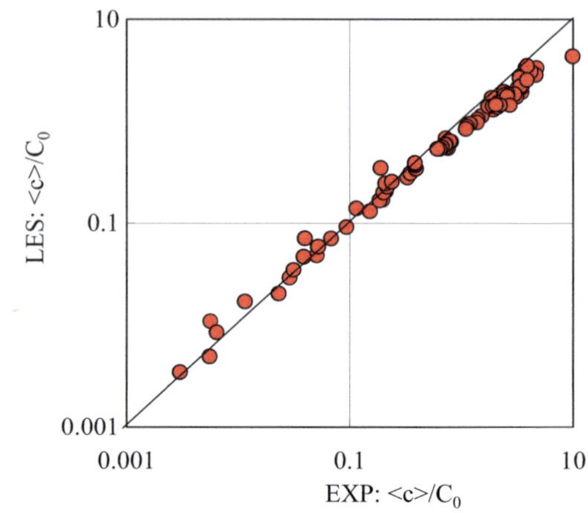

Fig. 9.32 Correlation between measured concentration and calculated concentration by LES

9.5 Conclusion

CFD is very promising for predicting and assessing air pollution and heat island phenomena in urban areas. Both phenomena have become serious problem especially in Asian countries with the rapid advance of urbanization. In order to appropriately apply CFD techniques to estimate ventilation and pollutant/thermal dispersion in urban areas, it is indispensable to validate CFD by comparing calculated results with reliable experimental data. In this chapter, a wind tunnel experimental technique for gas/thermal dispersion was explained, and some results of validation studies on large eddy simulation of pollutant/thermal dispersion in non-isothermal boundary layer were introduced.

Acknowledgments The author is greatly indebted to his postdoctoral researchers, PhD students, and master students for developing the measuring technique and conducting wind tunnel experiments and CFD simulations included in this chapter. Related members of this chapter are as follows.

Hideyuki Tanaka, Taich Shirasawa, Guoyi Jiang, Tingting Hu, Tsuyoshi Kobayashi, Koudai Katada, and Keisuke, Nomura.

References

Abe S, Tamura T, Nakayama H (2008) LES on turbulence structures and gaseous dispersion behavior in convective boundary layer with the capping inversion. J Wind Eng 33

Architectural Institute of Japan (2007) Guide book for practical applications of CFD to pedestrian wind environment in urban areas, Maruzen Company, Limited

Brillant G, Husson S, Bataille F, Ducros F (2008) Study of the blowing impact on a hot turbulent boundary layer using thermal large eddy simulation. Int J Heat Fluid Flow 29:1670–1678

Druault P, Lardeau S, Bonnet JP, Coiffet F, Delville J, Lamballais E, Largeau JF, Perret L (2004) Generation of three-dimensional turbulent inlet conditions for large eddy simulation. AIAA J 42(3):447–456
Hattori H, Houra T, Nagano Y (2007) Direct numerical simulation of stable and unstable turbulent thermal boundary layers. Int J Heat Fluid Flow 28:1262–1271
Iizuka S, Murakami S, Tsuchiya N, Mochida A (1999) LES of flow past 2D cylinder with imposed inflow turbulence. In: Proceedings of 10th international conference on wind engineering 2, Copenhagen, pp 1291–1298
ISO (1993) Guide to the expression of uncertainty in measurement, Genève
Jiang G, Yoshie R, Shirasawa T, Jin X (2012) Inflow turbulence generation for large eddy simulation in non-isothermal boundary layers. J Wind Eng Ind Aerodyn 104–106:369–378
Kataoka H (2008) Numerical simulations of a wind-induced vibrating square cylinder within turbulent boundary layer. J Wind Eng Ind Aerodyn 96:1985–1997
Kataoka H, Mizuno M (2002) Numerical flow computation around aeroelastic 3D square cylinder using inflow turbulence. Wind Struct 5:379–392
Kondo K, Murakami S, Mochida A (1997) Generation of velocity fluctuations for inflow boundary condition of LES. J Wind Eng Ind Aerodyn 67/68:51–64
Kong H, Choi H, Lee JS (2000) Direct numerical simulation of turbulent thermal boundary layers. Phys Fluids 12:2555–2568
Lund TS, Wu XH, Squires KD (1998) Generation of turbulent inflow data for spatially-developing boundary layer simulations. J Comput Phys 140:233–258
Maruyama T, Rodi W, Maruyama Y, Hiraoka H (1999) Large eddy simulation of the turbulent boundary layer behind roughness elements using an artificially generated inflow. J Wind Eng Ind Aerodyn 83:381–392
Nagano Y, Kim C (1988) A two-equation model for heat transport in wall turbulent shear flows. Trans ASME, J Heat Transf 110:583–589
Nozawa K, Tamura T (2002) Large eddy simulation of the flow around a low-rise building immersed in a rough-wall turbulent boundary layer. J Wind Eng Ind Aerodyn 90:1151–1162
Nozawa K, Tamura T (2005) Large eddy simulation of wind flows over large roughness elements. In: Proceedings of EACWE4, Prague, pp 1–6
Ohya Y (2001) Wind-tunnel study of atmospheric stable boundary layers over a rough Surface. Bound-Lay Meteorol 98:57–82
Ohya Y, Uchida T (2004) Laboratory and numerical studies of the convective boundary layer capped by a strong inversion. Bound-Lay Meteorol 112:223–240
Ohya Y, Uchida T (2008) Laboratory and numerical studies of the atmospheric stable boundary layers. J Wind Eng Ind Aerodyn 96:2150–2160
Okaze T, Mochida A (2014) A generation method for turbulent fluctuation of wind velocity and scalar based on Cholesky decomposition of turbulent fluxes, Generation of inflow turbulence including scalar fluctuation for LES Part 1. J Environ Eng (Trans AIJ) 79(703):771–776
Perret L, Delville J, Manceau R, Bonnet JP (2006) Generation of turbulent inflow conditions for large eddy simulation from stereoscopic PIV measurements. Int J Heat Fluid Flow 27:576–584
Tabor GR, Baha-Ahmadi MH (2010) Inlet conditions for large eddy simulation: a review. Comput Fluids 39:553–567
Tamura T (2008) Towards practical use of LES in wind engineering. J Wind Eng Ind Aerodyn 96:1451–1471
Tamura T, Tsubokura M, Cao S, Furusawa T (2003) LES of spatially-developing stable/unstable stratified turbulent boundary layers. Direct Large Eddy Simul 5:65–66
Tamura T, Nakayama J, Ohta K, Takemi T, Okuda Y (2006) LES estimation of environmental degradation at the urban heat island due to densely-arrayed tall buildings. In: 17th symposium on boundary layers and turbulence, AMS, San Diego, pp 22–25
Tominaga Y, Mochida A, Yoshie R, Kataoka H, Nozu T, Yoshikawa M, Shirasawa T (2008a) AIJ guidelines for practical applications of CFD to pedestrian wind environment around buildings. J Wind Eng Ind Aerodyn 96:1749–1761

Tominaga Y, Mochida A, Murakami S, Sawaki S (2008b) Comparison of various revised k-ε models and LES applied to flow around a high-rise building model with 1:1:2 shape placed within the surface boundary layer. J Wind Eng Ind Aerodyn 96:389–411
Xie ZT, Castro IP (2008) Efficient generation of inflow conditions for large eddy simulation of street-scale flows. Flow, Turbul Combust 81:449–470
Yoshie R, Hu T (2013) Effect of atmospheric stability on urban pollutant concentration. In: Proceedings of the eighth Asia-Pacific conference on wind engineering, Chennai
Yoshie R, Mochida A, Tominaga Y, Kataoka H, Harimoto K, Nozu T, Shirasawa T (2007a) Cooperative project for CFD prediction of pedestrian wind environment in Architectural Institute of Japan. J Wind Eng Ind Aerodyn 95:1551–1578
Yoshie R, Tanaka H, Shirasawa T (2007b) Technique for simultaneously measuring fluctuating velocity, temperature and concentration in non-isothermal flow. In: Proceedings of the 12th International conference on wind engineering, Cairns, pp 1399–1406
Yoshie R, Jiang GY, Shirasawa T, Chung J (2011) CFD simulations of gas dispersion around high-rise building in non-isothermal boundary layer. J Wind Eng Ind Aerodyn 99:279–288

MIX
Papier aus verantwortungsvollen Quellen
Paper from responsible sources
FSC® C105338

If you have any concerns about our products,
you can contact us on
ProductSafety@springernature.com

In case Publisher is established outside the EU,
the EU authorized representative is:
Springer Nature Customer Service Center GmbH
Europaplatz 3, 69115 Heidelberg, Germany

Printed by Libri Plureos GmbH
in Hamburg, Germany